T0332554

Nanoscale Technologies in Plant Sciences
PRINCIPLES AND APPLICATIONS

Nadeesh M. Adassooriya

Department of Chemical and Process Engineering
Faculty of Engineering, University of Peradeniya
Peradeniya, Sri Lanka

Ryan Rienzie

Department of Agricultural Biology
Faculty of Agriculture, University of Peradeniya
Peradeniya, Sri Lanka

Nadun H. Madanayake

Department of Botany
Faculty of Science, University of Peradeniya
Peradeniya, Sri Lanka

CRC Press
Taylor & Francis Group
Boca Raton London New York

CRC Press is an imprint of the
Taylor & Francis Group, an **informa** business

First edition published 2023
by CRC Press
6000 Broken Sound Parkway NW, Suite 300, Boca Raton, FL 33487-2742

and by CRC Press
4 Park Square, Milton Park, Abingdon, Oxon, OX14 4RN

© 2023 Nadeesh M. Adassooriya, Ryan Rienzie and Nadun H. Madanayake

CRC Press is an imprint of Taylor & Francis Group, LLC

Library of Congress Cataloging-in-Publication Data (applied for)

ISBN: 978-1-032-38792-5 (hbk)
ISBN: 978-1-032-38806-9 (pbk)
ISBN: 978-1-003-34685-2 (ebk)

DOI: 10.1201/9781003346852

Typeset in Times New Roman
by Innovative Processors

Preface

With the advancement of nanotechnology, nanoscale materials used in industrial applications have been exponentially increasing which triggers the release of nanomaterials into the environment deliberately or accidentally. These materials tend to get absorbed by plants which are the primary producers in most trophic levels. Nanomaterials can influence plant metabolisms causing positive effects, improving plant growth and development or negative effects such as phytotoxicity. These impacts are dependent on the properties of nanoparticles, mode of interaction and concentration as well as the environmental conditions. Therefore, understanding the nanoscale technologies in plant sciences along with the fundamentals behind the behavior of nanoscale materials is timely and important.

The primary focus of this book is the interactions of nanomaterials with plants while discussing the fundamentals of nanotechnology. The book provides a comprehensive knowledge of plant and nanomaterials interactions, nanodelivery systems present in agriculture, impacts of nanomaterials on seed germination, nanosensors for plant disease diagnosis and toxicity aspects of nanomaterials towards plants. In addition, the fundamentals of nanotechnology including the beauty of the nanoscale, nanomaterials synthesis and more importantly nanomaterials characterization techniques are discussed. Furthermore, special attention has been given to converse practical aspects related to the aforementioned subsections. This will help the researchers transform their knowledge into real-life applications on different scales.

Nanotechnology has evolved in the literature bias toward chemistry and physics which creates a barrier for biologists to pursue nanosciences. Compiling textbooks focusing on plants and nanomaterials will assist to narrow down those walls which inculcate confidence among biologists who wants to take part in adventures in the emerging field of nanotechnology.

We hope this book will be useful to undergraduate and postgraduate level researchers, academics, educators and advanced researchers who are concerned about the interactions of nanomaterials with plants. We also hope that the efforts to forward the readers toward a better understanding of nanomaterial and plant interactions shall be fruitful.

Nadeesh M. Adassooriya
Ryan Rienzie
Nadun H. Madanayake

Contents

Overview and Introduction to Nanotechnology

Development in science and technology rapidly envisioned the use of novel and efficient technologies in different applications. Nanotechnology has emerged as one such rapidly advancing technology in the recent past. Nanotechnology can be defined as the science of designing, characterizing, manufacturing and application of materials on the nanoscale with controlled sizes and shapes. Nanomaterials are highly reactive moieties that are utilized in nanotechnology-based interdisciplinary approaches such as in food, agriculture, medicine, energy, cosmetics, environmental remediation, etc. However, the reactivity of nanomaterials directly interact with biological systems including flora and fauna, and microorganisms. These interactions have pros as well as cons on living components. As nanomaterials are released into the environment researchers have shown a significant interest to assess their potential impacts on the biosphere. Plants are the primary producers in many of the ecosystems in the world. Nanomaterials can interact directly or indirectly with plants which might lead to trophic transfer influencing other trophics (Madanayake et al. 2022). The application of nanoparticles as fertilizers, pesticides or growth promoters can directly interact with plants. Sometimes nanomaterials released from industries or during environmental remediation strategies can indirectly interact with plants. Therefore, understanding the fate and behavior of nanomaterials in the environment and their impact on the plants is imperative. Hence, this book provides an overview of nanotechnology, synthesis, and characterization of nanomaterials. Also, it covers nanomaterials' interactions with plants concerning their intake and translocation in the vegetative system, impacts on seed germination and growth as well as on toxicity of nanomaterials on plants. Moreover, this book elaborates on the nanodelivery systems for quality enrichment in plants and nanosensors for plant disease diagnosis.

1.1 Nanoscale and History of Nanotechnology

Materials that we utilize at the nanoscale can be named nanoparticles. Nanoparticles are materials with at least a single dimension in the range of 1–100 nm. Nanotechnology can be named as an interdisciplinary approach to implementation in all fields of science and technology (see Figure 1.1). Materials at this scale show unique properties compared to their bulk states. When it comes to this scale (below 100 nm) they start to act differently. This is imperative for numerous applications for several reasons. Although this is a novel technique to the world, it prevails from the time of the beginning of life on earth. In nature, this applies in the synthesis of biologically important components such as proteins, enzymes, carbohydrates, deoxyribose nucleic acids (DNA), ribose nucleic acids (RNA), viruses and many others. Therefore, this has become a promising field to develop different products for life

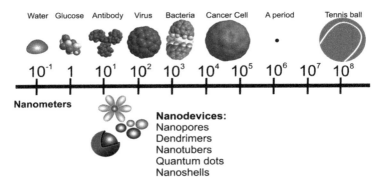

Figure 1.1: Understanding the nanoscale. Nanoparticles are materials with at least one of their dimensions found within the 1–100 nm range.

science applications. Hence, scientists are trying to master the art of this technology to develop environmentally sustainable products.

Nanotechnology goes back to far history and is revealed through analyzing various crafts and equipment produced in various eras, some exciting examples from different timescales are given below.

- During the 4th Century: The Lycurgus Cup of Rome which is a dichroic glass with colloidal gold and silver in the glass. The glass becomes green when it is lit from the outside, while translucent red when light shines through the inside.
- In the 6th–15th Centuries: In European Churches, stained glass windows became popular with gold nanoparticles, other metal oxides and chlorides
- In the 9th–17th Centuries: Glowing, glittering ceramic glazes were used in Islamic religious places and later the technology was adopted by Europeans too.
- From the 13th–18th Centuries: It was found to contain carbon nanotubes and cementite nanowires in "Damascus" saber blades (Bayda et al. 2019).

1.2 Modern Nanotechnology

The foundation for modern nanotechnology was laid in the 1800s and the attention of the scientific community for modern nanotechnology was laid with the concept of "plenty of space at the bottom" by the American physicist Richard Feynman in 1959 (see Figure 1.2). However, this concept was less popular in the scientific community until the 1980s (Bayda et al. 2019).

Figure 1.2: Richard Feynman (https://press.princeton.edu).

Following is a series of events that includes the introduction of the concept "plenty of space at the bottom" by Richard Feynman.

1857: Michael Faraday's discovery of colloidal gold

Development of new instruments that aided characterization

1936: Erwin Müller invented the field emission microscope

1950: Victor La Mer and Robert Dinegar developed the theory and a process for growing monodisperse colloidal materials

1951: Erwin Müller: field ion microscope to image the arrangement of atoms

1956: Arthur von Hippel introduced the concept of "molecular engineering"

1959: Richard Feynman's concept "There's Plenty of Room at the Bottom"

1965: Gordon Moore laid the foundation for an electronic transition era

1974: Norio Taniguchi introduced the term nanotechnology

1981: Gerd Binnig and Heinrich Rohrer at IBM's Zurich lab invented the scanning tunneling microscope. They won the Nobel Prize for this discovery in 1986

1981: Alexei Ekimov from Russia discovered nanocrystalline, semiconducting quantum dots in a glass matrix

1985: Discovery of fullerene by Sir Harold W. Kroto, Richard E. Smalley and Robert F. Curl, Jr. who won the Nobel Prize for Chemistry in 1996

1985: Discovery of colloidal semiconductor nanocrystals (quantum dots) by Louis Brus

1986: Gerd Binnig, Calvin Quate, and Christoph Gerber invented the atomic force microscope

1989: Don Eigler and Erhard Schweizer at IBM demonstrated precise manipulation of atoms

The 1990s: Nanotechnology became an industry

1991: Carbon nanotubes were discovered by Japanese scientist Iijima Sumio

1992: C.T. Kresge and colleagues discovered mesoporous silica-based nanostructured catalytic materials

1993: Moungi Bawendi invented a method for controlled synthesis of nanocrystals (quantum dots)

1999: Wilson Ho and Hyojune Lee probed secrets of chemical bonding by assembling a molecule from Fe and carbon monoxide (CO) with a scanning tunneling microscope

1999: Chad Mirkin invented dip-pen nanolithography

1999–early 2000's: Consumer products with nanotechnology came to the market

2000-2014: Strategizing and policing concerning issues associated with nanotechnology by UK and USA

2003: Naomi Halas, Jennifer West, Rebekah Drezek, and Renata Pasqualin developed gold nanoshells

2005: Erik Winfree and Paul Rothemund developed theories that help in nanocrystal growth

2007: Angela Belcher and colleagues built a lithium-ion battery

2009–2010: Nadrian Seeman and colleagues created DNA-like robotic nanoscale assembly devices

2010: IBM demonstrated nanoscale patterns and structures as small as 15 nanometers using a Si tip

(Nano.gov 2009)

1.3 What are Nanomaterials?

Nanomaterials are defined as materials possessing, at minimum, one external dimension measuring 1–100 nm. (Bhushan 2017). In general, nanomaterials can be categorized into four types namely zero-dimensional (nanoparticles) single dimensional (nanorods and nanotubes), two-dimensional (nanosheets) and three-dimensional (graphite) (Khan 2020) and these types can be further divided into three major categories namely, natural, incidental, and engineered. Figure 1.3 elaborates further on the classification of nanomaterials based on its dimensionality. Nanomaterials are generally synthesized using top-down and bottom-up approaches. Top-down approaches are used to synthesize nanomaterials by breaking down materials at the macroscale into nanosized particles. In addition, the bottom-up approach fabricates nanomaterials via aggregation or nucleation of atoms to form nanosized materials (Tulinski and Jurczyk 2017). These fabrication techniques will be discussed in detail in Chapter 2 of this book.

Nanostructures

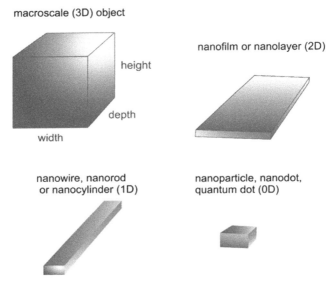

Figure 1.3: Classification of nanostructures based on their dimensionality.

One of the best examples to classify nanostructures based on their dimensionality is carbon-based nanomaterials (see Figure 1.4). Fullerene is a 0D nanomaterial where all its dimensions are restricted within a 1–100 nm scale. Carbon nanotubes are 1D nanostructures where only two dimensions lie within the nanoscale while one of its

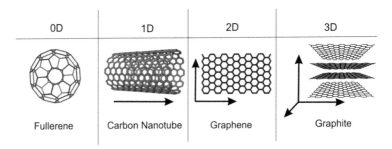

Figure 1.4: Different carbon-based nanomaterials are grouped based on their dimensionality.

dimensions exceeds the nanoscale (simply beyond 100 nm). Graphene is a single layer of graphite which is a 2D carbon-based nanomaterial which includes only a single dimension to retain within the nanoscale. Stacked layers of graphene form graphite and this can be defined as a 3D nanomaterial in which all of its dimensions are beyond 100 nm scale.

1.4   Why Nanomaterials are Unique?

At the nanoscale, the most desirable properties can be achieved by the dominance of size, distribution, morphology, and quantum effects. Physicochemical properties of nanomaterials distinctly alter with respect to single atoms and bulk counterparts for material with the same chemical composition. The higher surface area to volume ratio of nanomaterials makes them highly reactive moieties. This enhances the fraction of molecules or atoms at the surface of a given nanomaterial. This is ideal to be used as catalysts. Mostly, nanomaterials show size-dependent unique properties. When the size of a given material diminishes into the nanoscale, the surface area to volume ratio significantly increases. Once this happens the fraction of molecules or atoms that comes to the surface of a given material gets increased. The increase in the number of surface species enhances the potential of nanomaterials to interact with other reactant species. Furthermore, atoms in the interior of the nanomaterial are highly coordinated. However, the coordination number for the atoms at the surface of these nanomaterials is less coordinated compared to inner ones. Therefore, surface atoms are free to coordinate or make bonds with external molecules thus showing their higher reactivity. This is imperative because this enhances their catalytic capability. Hence, nanomaterials are highly reactive when their size reduces. In addition, lower coordination decreases the stability of the surface atoms of nanomaterials (Nasrollahzadeh et al. 2019). This may lead to altering their physical properties as well. Hence, a higher surface area to volume ratio improves the reactivity of nanoparticles.

Also, quantum confinement of nanomaterials alters their electronic and optical properties and this diversifies their potential for different applications than their bulk counterparts (Abou El-Nour et al. 2010). Optical properties of the nanomaterial including absorption, transmission, reflection, and light emission alter significantly. For example, gold at bulk state appears in yellow or golden color whereas the color of gold nanoparticles can show variable colors other than yellow. These optical effects can be manipulated based on their shape, size, and surface functionality. Optical properties of nanomaterials are capable of confining their electrical properties to produce a quantum effect. It can be simply explained as follows, monodisperse gold nanoparticles (~30 nm) can absorb light in the blue-green portion of the visible spectrum (~450 nm) and they reflect red light (~700 nm) from the visible spectrum yielding red color. As the particle size increases, the absorption shifts towards longer wavelengths. In other words, gold nanoparticles will absorb red light and blue light is reflected, yielding solutions with a pale blue or purple color as the size increases. As particle size continues to increase toward the bulk state absorption, wavelengths of gold will shift into the infrared portion of the spectrum. This will start to reflect the most visible wavelengths giving the nanoparticles a clear or translucent color. Once light rays are absorbed by a nanoparticle, oscillating electric fields of a light ray start to interact with the free electrons at the surface. This results in the vibration of electrons that is in resonance with the frequency of visible light and thus interacted. These resonant oscillations are known as surface plasmon. The colors shown by different nanomaterials are due to this phenomenon and it is known as Surface Plasmon Resonance. The surface plasmon resonance can be fine-tuned by altering the size and shape of the nanoparticles. Furthermore, the optical properties shown by nanomaterials differ from their chemical composition. For example, spherical-shaped gold nanomaterial at 100 nm diameter appears orange while silver nanoparticles with the same dimensions appear yellow in color.

The electronic properties of nanomaterials are also unique when compared to their bulk materials. Electronic band structure can be used to explain the electronic properties of nanomaterials. This depends on the particle size. When the size of nanoparticles decreases, the band gap of the conduction and valence band energy levels increases and does not exceed the band gap energies related to insulators. Therefore, the metallic character of metallic nanoparticles gradually changes to the semiconductor nature. This shows that nanomaterials demonstrate size and shape-dependence allowing them to be used in numerous applications.

1.5 Types of Nanomaterials

Nanomaterials are being manufactured for various applications. Based on their composition these can be classified as carbon-based nanomaterials, metal-based nanomaterials, metal oxides and metal salts, quantum dots, and polymeric nanomaterials (see Figure 1.7. Type-wise they are unique in their properties and understanding them is important as a nanotechnologist. Also, there are naturally available nanomaterials in the environment as well.

1.5.1 Natural Nanomaterials

Nanomaterials associated with natural systems come under this category. Perhaps one may have experienced the two wonders of the water repelling effect of lotus leaves and the geckos running on the ceilings or vertical glasses without falling in their childhood. The lotus leaves do not become wet upon water collecting on them, because the lotus leaves have millions of tubules or nanostructures made out of wax that hold water and with a very small shock the water is removed from the leaf surface (see Figure 1.6, Costa et al. 2013). Hundreds of nanosized hairs present on the bottom of the feet of gecko's feet (see Figure 1.5) help them to adhere to the ceiling or glass surfaces without falling onto the ground.

Figure 1.5: Structure of gecko's feet. Source: http://sustainable-nano.com.

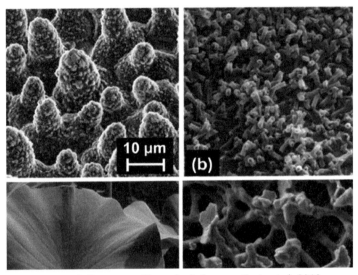

Figure 1.6: Structure of lotus leaf. Source: Costa et al. 2013.

1.5.2 Incidental Nanomaterials

Such nanomaterials are produced as byproducts of various processes that mainly include combustion processes. For instance, vehicle engines, produce nanoparticles and emit them into the environment. Welding fumes also contain nanomaterials. In our day to day life, burning a piece of paper can lead to producing nanomaterials (carbon nanotubes).

1.5.3 Engineered Nanomaterials

These nanomaterials are produced by man for desired applications such as in agriculture, medicine or other industries. They can be categorized into several categories as follows.

Carbon-based Nanomaterials

Carbon-based nanomaterials are nanoproducts that are in great demand in the industrial sectors. This includes carbon nanotubes, fullerenes, graphene carbon nanofibers, carbon black, and carbon onions. Carbon nanotubes are rolled sheets of graphene. Mostly these are utilized in structural reinforcement as they are 100 times stronger

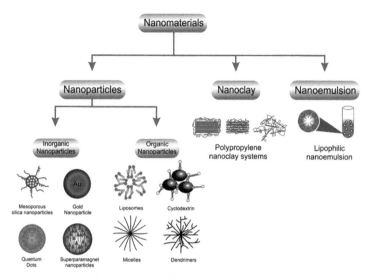

Figure 1.7: Engineered nanomaterials.

than steel. In addition, carbon nanotubes can be further classified as single-walled carbon nanotubes and multi-walled carbon nanotubes (see Figure 1.8). Single-walled nanotubes are comprised of a single graphite sheet wrapped into a cylindrical tube, while multi-walled nanotubes are built up with many concentric rings. These nanotubes are unique because of their thermal conductivity. They are conductive and non-conductive across the length of the tube.

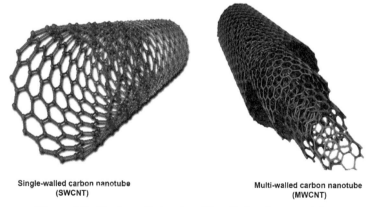

Figure 1.8: Single-walled and multi-walled carbon nanotubes.

Graphene consists of C atoms, each is covalently bonded to three others with angles of 120° and arranged as a single layer like a chicken wire or honeycomb. Also, oxides of graphene are another form of carbon-based nanomaterial. Graphene oxide's biodistribution and kinetic properties can be altered by controlling their size and surface chemistry, while increase the tensile strength which is a measure of the ability of a material to withstand the stress that is applied longitudinally without breaking. This can be increased by incorporating other materials.

Fullerenes are the allotropes of carbon having a structure of a hollow cage of sixty or more carbon atoms. Fullerene is comprised of 60 carbon atoms with sp^2 hybridization status, arranged spherically as 12 closed pentagons and 20 hexagons. The structure of carbon is called Buckminsterfullerene because it looks like a hollow football. The carbon units in these structures have pentagonal and hexagonal arrangements. These have commercial applications due to their electrical conductivity, structure, strength, and electron affinity. Figure 1.9 pictorially represents the structures of graphene and fullerene.

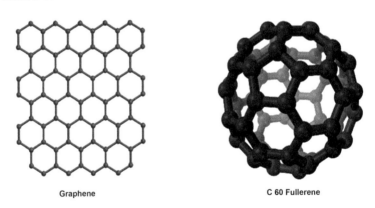

Graphene   C 60 Fullerene

Figure 1.9: Graphene and fullerene.

Metallic Nanoparticles

Nanoparticles synthesized using metal and metal oxide precursors are known as metal or metal oxide nanoparticles. Metallic nanoparticles are the most flexible of nanostructures due to their ability to control their physical and chemical properties. Such properties were mentioned previously in this chapter of the book. Moreover, metallic nanoparticles are easy to prepare. Even with ordinary laboratory facilities, metallic nanoparticles are easy to prepare using normally available reagents and it is possible even with plant extracts. Ag, Cu, and Mn can form nano forms, viz Ag, Cu and Mn. Gold (Au), nickel (Ni), silver (Ag), platinum (Pt), iron (II, III) oxide (Fe_2O_3, Fe_3O_4, FeO), zinc oxide (ZnO), gadolinium (Gd), and titanium dioxide (TiO_2) nanoparticles are some of the examples for this category. The limited size and higher number of protruding sites such as corners or edges on the surface make them unique, owing physical and chemical properties that are unique to metallic nanoparticles in general. These nanoparticles have applications in the detection and imaging of biomolecules and environmental and bioanalytical applications (Falcaro et al. 2016). For instance, Au nanoparticles are used to coat the biological sample (gold sputtering) before they are analyzed using electron microscopic techniques. In addition, iron-based nanomaterials are used as magnetic nanoparticles in nanotheranostics applications for targeted drug delivery and imaging (Madanayake et al. 2019). Therefore, these are one of the most imperative nanomaterials used in industrial applications.

Polymeric Nanomaterials

Polymeric nanomaterials are organic-based nanoparticles where numerous polymeric monomers are used in the synthesis. Both natural and synthetic polymeric precursors are used to synthesize polymeric nanoparticles. Poly (lactide), poly (lactide-co-glycolide) copolymers, poly (ε-caprolactone), poly (amino acids) and natural polymers like alginate, chitosan, gelatin, and albumin are precursors for the production. Based on the synthesis methods these materials can have different shapes such as nanocapsules or nanospheres or nanofibres etc. These are utilized in controlled release drugs and bioactive compounds, protection of drug molecules, therapeutics, and targeted drug deliveries (Masood 2016, Bhargava 2018).

Quantum Dots

Quantum dots are semiconductor nanocrystals with unique optical and electronic properties. The properties of quantum dots are size and composition-dependent (Rafique et al. 2020). Quantum dots can emit light with various colors once it interacts with UV light. Quantum dots have a crystalline structure and they can transport electrons. When the particles contact ultraviolet light, they emit light in various colors. When the size of the quantum dots get decreased, the quantum effects become more prominent. In other words, smaller quantum dots can effectively emit or absorb energy in the visible spectrum. These semiconducting nanoparticles are used as composites in solar cells and as fluorescent biological labels and in electronic and biomedical applications. For example, the electronic properties of quantum dots allow them to be used as active materials in single-electron transistors. Quantum dots can convert a spectrum of light into different colors and each dot emits a different color depending on its size. Furthermore, these nanocrystals show fluorescence at different wavelengths depending on the size. Therefore, quantum dots make them promising candidates for fabricating optical probes for biological and medical imaging (Geszke-Moritz et al. 2013). For instance, they can be effectively used in developing biological sensors to detect various chemicals. For example, to identify pesticide residues in food, and identify pathogens and pests using the specific chemicals emitted by them (Jiang et al. 2019).

1.6   Conclusion and Future Perspectives

The unique and inherent properties of nanomaterials relating to their bulk counterparts have enabled the optimization of the efficiency of different industrial applications. Different types of nanomaterials including carbon-based nanomaterials, metal and metal oxide nanoparticles, polymeric nanomaterials and quantum dots are some of the prominently used materials in nano-enabled products. They have been widely used in food, agriculture, medicine, energy, cosmetics, environmental remediation, electronics, etc. Even though they have been utilized in different applications, higher reactivity nanomaterials have a significant impact on the environment. Therefore, understanding the fate and behavior of nanomaterials in the environment and their toxic impact is of utter importance to optimize the advantages of this technique to the world.

References

Abou El-Nour, K.M., A.A. Eftaiha, A. Al-Warthan and R.A. Ammar. 2010. Synthesis and applications of silver nanoparticles. *Arabian Journal of Chemistry*, 3(3), pp. 135–140.

Bayda, S., M. Adeel, T. Tuccinardi, M. Cordani and F. Rizzolio. 2019. The history of nanoscience and nanotechnology: From chemical–physical applications to nanomedicine. *Molecules*, 25(1), p. 112.

Bhargava, S. 2018. *Polymeric Nanomaterials and Their Applications.* Doctoral dissertation, National University of Singapore (Singapore).

Bhushan, B. 2017. Introduction to nanotechnology. In: *Springer Handbook of Nanotechnology* (pp. 1–19). Springer, Berlin, Heidelberg.

Costa, M.N., B. Veigas, J.M. Jacob, D.S. Santos, J. Gomes, P.V. Baptista, R. Martins, J. Inácio and E. Fortunato. 2014. A low-cost, safe, disposable, rapid and self-sustainable paper-based platform for diagnostic testing: Lab-on-paper. *Nanotechnology*, 25(9), 094006.

Falcaro, P., R. Ricco, A. Yazdi, I. Imaz, S. Furukawa, D. Maspoch, R. Ameloot, J.D. Evans and C.J. Doonan. 2016. Application of metal and metal oxide nanoparticles@ MOFs. *Coordination Chemistry Reviews*, 307, pp. 237–254.

Geszke-Moritz, M. and M. Moritz. 2013. Quantum dots as versatile probes in medical sciences: Synthesis, modification and properties. *Materials Science and Engineering: C*, 33(3), pp. 1008–1021.

Jiang, M., J. He, J. Gong, H. Gao and Z. Xu. 2019. Development of a quantum dot-labelled biomimetic fluorescence immunoassay for the simultaneous determination of three organophosphorus pesticide residues in agricultural products. *Food and Agricultural Immunology*, 30(1), pp. 248–261.

Khan, F.A. 2020. Nanomaterials: Types, classifications, and sources. In: *Applications of Nanomaterials in Human Health* (pp. 1–13). Springer, Singapore.

Madanayake, N.H., N. Perera and N.M. Adassooriya. 2022. Engineered nanomaterials: Threats, releases, and concentrations in the environment. In: *Emerging Contaminants in the Environment* (pp. 225–240). Elsevier.

Madanayake, N.H., R. Rienzie and N.M. Adassooriya. 2019. Nanoparticles in nanotheranostics applications. In: *Nanotheranostics* (pp. 19–40). Springer, Cham.

Masood, F. 2016. Polymeric nanoparticles for targeted drug delivery system for cancer therapy. *Materials Science and Engineering: C*, 60, pp. 569–578.

Nano.gov. 2009. *Nanotechnology Timeline | Nano.* [online] Available at: https://www.nano.gov/timeline.

Nasrollahzadeh, M., S.M. Sajadi, M. Sajjadi and Z. Issaabadi. 2019. An introduction to nanotechnology. In: *Interface Science and Technology* (Vol. 28, pp. 1–27). Elsevier.

Rafique, M., M.B. Tahir, M.S. Rafique, N. Safdar and R. Tahir. 2020. Nanostructure materials and their classification by dimensionality. In: *Nanotechnology and Photocatalysis for Environmental Applications* (pp. 27–44). Elsevier.

Tulinski, M. and M. Jurczyk. 2017. Nanomaterials synthesis methods. *Metrology and Standardization of Nanotechnology: Protocols and Industrial Innovations*, pp. 75–98.

Synthesis of Nanomaterials

Nanomaterial synthesis has gained marked attention because it is important to synthesize materials with uniformity and the desired properties for the application of interest. Different nanomaterials including metallic nanoparticles with magnetic properties are used in targeted drug delivery in theranostic applications (Madanayake et al. 2019). In addition, in agricultural applications scientists have focused on utilizing different nanomaterials for controlled and slow release of plant nutrients. Sometimes nanoparticles themselves can be used as a nutrient source. Furthermore, nanoparticles such as zero-valent iron for the elimination of environmental pollutants and carbon nanotubes for applications which require high strength, durability, electrical conductivity etc. Also, some nanomaterials as molecular electronics, sensing devices or reinforcing additive fibres in functional composite materials. Therefore, different applications require different nanoparticles native to the application.

Several different techniques have been proposed to synthesize nanomaterials and these strategies can be grouped as top-down or bottom-up approaches (Baig et al. 2021) and Figure 2.1 graphically represent these approaches of synthesizing nanomaterials. In the top-down approach, bulk materials are broken down into nanosized units using a source of energy i.e., chemical, mechanical etc. Mechanical grinding through ball milling, chemical etching and lithography are typical examples of the methods involved in the synthesis of nanoparticles via a top-down approach. These techniques are simpler and are based on the division of bulk material to produce the desired structure with appropriate properties. Primarily, this approach works on two parameters: first is creating nanostructures by "sculpting" and secondly, by creating nanostructures by adding or rearranging the desired material onto the substrate. Sculpting includes etching (dry or wet), mechanical attrition, and lithography techniques which eliminates undesired material to bring in the desired shape. In contrast, bottom-up approaches, involve the self-assembling of materials either molecule by molecule or atom by atom or even nanosized particles into nanoparticles with the desired size through chemical or physical reactions. Chemical and physical vapour deposition, sol-gel, sputtering, epitaxy, evaporation-condensation and spray pyrolysis techniques are examples of the methods involved with a bottom-up approach. Bottom-up approaches can be highly precise in sizes, shapes and structures for scale-up production. Sometimes, bottom-up approaches are promising compared to top-down methods because top-down approaches might result in imperfections in the surface structure of nanoparticles. This would have a significant impact on the physical properties and the surface chemistry of the nanostructures and the nanomaterials thus synthesized.

Methods involved in nanomaterial synthesis can be physical, chemical or biological methods. Physical and chemical-based techniques are the most widely explored and conventional methods of nanomaterials fabrication. Recently, the interest in biological methods has gained greater attention due to the simplicity, eco-friendliness and due to lower toxicity of the biological components used in the synthesis. Hence, in this chapter, we will highlight some

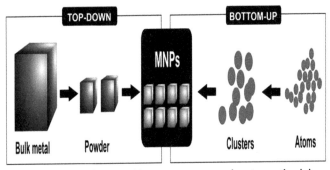

Figure 2.1: Top-down and bottom-up approaches to synthesizing metallic nanoparticles.

of the commonly used physical, chemical and biological methods used in nanomaterial synthesis.

2.1 Chemical Methods

2.1.1 Co-precipitation

This is one of the well-known methods of synthesizing nanomaterials and nanocomposites. This method of synthesis generally produces nanomaterials of insoluble species under higher supersaturation conditions which induce precipitation. Co-precipitation involves nucleation, growth, coarsening, and agglomeration of chemical precursors to form nanoparticles. Nucleation is the critical step in co-precipitation where several smaller particles are formed and this is followed by ripening. Ripening or Ostwald ripening is a phenomenon where smaller particles in a system dissolve and are deposited to form thermodynamically stable larger particles. Ripening and aggregation processes involved in the synthesis have a significant impact on the size, morphology, and physicochemical properties of nanomaterials.

$$A^+_{(aq)} + B^-_{(aq)} \rightarrow AB_{(s)}$$

Typically, co-precipitation is used in the synthesis of metal nanoparticles from aqueous solutions via chemical reduction, electrochemical reduction, and decomposing metallorganic precursors. Moreover, this has been used to synthesize metal oxides and metal chalconides as well (Rane et al. 2018, Kolahalam et al. 2019).

Nanoparticles in a solution can be considered as a colloidal suspension. Generally, solutions, colloids and suspensions are differentiated based on the size of the spread particles and their macroscopic properties. Simply colloids have 1–1000 nm-sized particles and nanoparticles are a subcategory within colloids. Stabilized nanoparticles in a suspension will not precipitate and they tend to show the Tyndall effect (see Figure 2.2). The Tyndall effect, also referred to as the Tyndall phenomenon is the scattering of a beam of light by a medium comprising of small suspended particles making the light beam entering the medium become visible. Therefore, this can be used as a simple confirmation test to check the successful synthesis of nanomaterials.

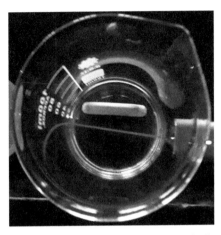

Figure 2.2: Visual observation of Tyndall effect of silver nanoparticles suspension synthesized by using sodium borohydride as the reducing agent.

2.1.2 Sol-gel Method of Nanomaterial Synthesis

This is a chemical procedure that has been widely applied in the fabrication of nanoparticles. It has been reported as a promising method to synthesize metal-oxide nanomaterials. In sol-gel synthesis, the reactants are converted to a sol leading to a network structure that is called a gel. Metal alkoxides are the prominent chemical precursors for the generation of nanomaterials using the sol-gel method. This process includes several steps and initially, hydrolysis of the metal oxide takes place in the aqueous phase or in the presence of alcohol to form a sol. Then condensation of the reaction mixture takes place leading to an increase in the solvent viscosity to formulate porous structures leaving to age. The condensation process leads to the formation of hydroxo or oxo bridges, forming metal hydroxo or metal oxo polymers in the solution and aging will lead to altering the structure, properties and porosity of the synthesized material. Then this follows a drying step to remove water and organic solvents from the gel, followed by calcination to obtain nanoparticles. Several factors affect the properties and the quality of the final product; reactants, hydrolysis rate, aging time, pH, and the molar ratio between H_2O and the precursor. This is a feasible method with many other benefits including higher homogeneity, lower processing temperature, and an ideal method to form composites and complex nanostructures (Rane et al. 2018, Baig et al. 2021).

2.1.3 Template Synthesis of Nanomaterials

Template synthesis methods are comprehensively used to produce nanoporous materials. Template synthesis can be classified as hard or soft template synthesis based on the template materials used. In the soft templating method, nanostructures are produced using block copolymers, flexible organic molecules, and anionic, cationic, and non-ionic surfactants as soft templates. Soft templates and the precursors interact with each other through hydrogen bonding, van der Waals forces, and electrostatic forces. The soft template method has been considered advantageous due to its straightforward implementation, relatively mild experimental conditions, and the development of materials with a range of morphologies. The hard template technique applies pre-designed solid materials as templates. Here, the porous structure of the hard template is filled up with reactant molecules to form nanostructures. It is imperative to maintain a mesoporous structure during the precursor conversion process, and they should be easily removable without disrupting the produced nanostructure. Different types of hard templates including carbon black, silica, carbon nanotubes, particles, colloidal crystals, and wood shells are used to synthesize nanomaterials. Template synthesis involves three main steps to synthesize nanomaterials. The first or foremost step is to select an appropriate template. Precursors are filled into the template and transformed into an inorganic solid. This is followed by template removal to obtain unique nanostructured materials such as nanowires, nanorods, 3D nanostructured materials, nanostructured metal oxides, and many other nanoparticles (Baig et al. 2021).

2.2 Physical Methods

2.2.1 Mechanical Attrition

The mechanical attrition processes help in producing non-crystalline alloys, ceramics, and composite nanostructures by milling tools under non-equilibrium conditions. Here, macro- or micro-sized particles grind in a ball mill, a planetary ball mill, or using any other type of size-reducing mechanism. Particles generated from these techniques are filtered and recovered for the desired application. Particle sizes of the synthesized materials can range from tens to hundreds of nanometers. The purity of the materials can be affected by contaminations from particles generated by mechanical tools during the operation. However, this can be mitigated by reducing the milling time as well as by utilizing pure and ductile raw materials.

2.2.2 Lithography

Nanolithography is a technique that is mainly used for the fabrication of nanostructures. This has been the key technology in the manufacturing of integrated circuits and microchips in the semiconductor industry. Lithography has advanced to techniques such as photolithography, electron beam lithography, focused ion beam lithography, soft lithography, nanoimprint lithography and scanning probe lithography mainly based on the energy source used. At present this has been used in electronics and microsystems, medical and biotechnology, optics and photonics, and environment and energy harvesting. This is a cost-effective and faster technique to fabricate nanomaterials using different types of radiation sources like photons, X-rays, electrons, ions and patterns that are transferred onto their respective resists (Baig et al. 2021).

Photolithography frequently utilizes in the semiconductor and integrated circuit industry. This has been applied for pattern generation to manufacturing integrated circuits, microchips and microelectromechanical systems devices. Photolithography utilizes a light-sensitive polymer known as photo-resist, as the irradiation source (UV) to define the desired pattern. Initially, UV light (193–436 nm) is illuminated through the photomask (e.g., quartz, glass) to make an exposure on a photo-resist coated onto a substrate. A photomask is an opaque plate with holes or transparencies that will allow light to shine in a defined pattern. Polymer components within the exposed area break and this will dissolve in the developer solution to remove the exposed photo-resist to form the desired photoresist pattern. This patterned photo-resist can be used as a protective layer for etching or deposition processes to build the topography on the substrate to synthesize nanomaterials.

Electron beam lithography can nanofabricate patterns by utilizing an accelerated electron beam focusing on an electron-sensitive resist to make an exposure. The diameter of this electron-beam spot ranges within a couple of nanometers where it scans the surface of resist in a dot-by-dot fashion to generate patterns in sequence. Instead of an electron beam, an accelerated ion beam is used in focused-ion beam lithography. Here an accelerated ion beam is directly punched onto a metallic film on the substrate. Focused ion beam systems are employed to deposit materials such as tungsten, platinum, and carbon. The resolution of electron beam and focused ion beam lithography techniques are in the order of 5–20 nm due to ultra-short wavelengths of electron/ion beams in the order of a few nanometers (Pimpin et al. 2012).

2.2.3 Gas Condensation

Gas condensation processes are used to synthesize crystalline metals and alloy nanoparticles. Here, a heat generating source such as joule heated refractory crucibles made of W, Ta or Mo are used to vaporize the precursor material. If the raw material tends to react with the crucibles, the electron beam evaporation technique can be used. Precursor materials used in the synthesis can be metal or an inorganic material. Vaporization of the materials needs to be performed at a pressure of 1–50 mbar. High residual gas pressure created in the system leads to the form of ultra-fine particles by gas phase collision. Clusters formed near the vaporization source take place by homogenous nucleation of the atoms in the gaseous phase. A nanocluster thus synthesized is collected using liquid nitrogen-filled cold finger scrapper assembly and a compaction device. The nanoparticles are removed by scrapper in the form of a metallic plate.

2.2.4 Chemical Vapour Deposition

This is one of the commonly used methods in synthesizing nanoparticles. In this process, a solid is deposited on a heated surface via a chemical reaction at the vapor phase. Chemical vapour deposition processes generally take place at higher temperatures (>900°C). Chemical vapour deposition systems comprise a gas supply, deposition chamber and exhaust system. The nanomaterials synthesized from this technique are usually deposited in the form of a thin film, powder, or as single crystals. Monitoring the reaction conditions of chemical vapour deposition aids in obtaining a range of nanomaterials with varying and controlled physical and chemical properties. This technique is widely used to manufacture dielectrics, conductors, passivation layers, oxidation barriers, conductive oxides and corrosion-resistant coatings, heat-resistant coatings, and epitaxial layers for microelectronics. Furthermore, this has been used for the synthesis of carbon-based nanomaterials especially carbon nanotubes (Carlsson and Martin 2010, Ijaz et al. 2020, Baig et al. 2021).

2.3   Biological Synthesis

The biological synthesis of nanomaterials includes the use of microorganisms, plant extracts, algae and their metabolites to synthesize nanomaterials. In this approach mostly microorganisms such as bacteria, fungi and algal species are used for the synthesis (Sundaram et al. 2012, Boroumand et al. 2015, Seetharaman et al. 2018, Pal et al. 2019). Microorganisms are regarded as eco-friendly green micro-factories. Therefore, microbial synthesis of nanomaterials is recognized as a promising arena which augments nanobiotechnology (Narayanan and Sakthivel 2010) and Table 2.1 shows some of the potential organisms that can be used to synthesize nanomaterials. Generally, living organisms produce different metabolites during their growth and development. Most of the compounds thus synthesized show reducing properties and this can be used to synthesize nanoparticles. Also, biological templates are used as major tools for synthesis. Biological templates produce unique and sophisticated nanostructures. This includes biological templates like DNA and proteins (Singh and Chakarvarti 2016). These can be used to design biosensors, bioelectronics systems etc. Proteins are the main constituents for the development of biological-based nanostructures as well. For instance, viral nanoparticles can be used in biomedical applications and these are regarded as prefabricated nanomaterials.

Plants and plant extracts have also been used for the synthesis of nanoparticles. Metal nanoparticles get reduced by the phytochemicals present in the plants. Phytochemicals like flavones, alkaloids and organic acids naturally act as better reducing agents for nanoparticle preparation. In algae, polysaccharides are harnessed to synthesize. Hydroxyl groups and functional groups of these molecules can act as reducing agents and agents to stabilize nanoparticles. Fungi utilize their intracellular and extracellular enzymes to produce nanomaterials whereas yeasts use membrane-bound oxidoreductase enzymes and quinones for the synthesis. Moreover, bacteria can synthesize nanomaterials by reducing metal ions using specific reducing enzymes like NADH-dependent reductase or nitrate-dependent reductases.

Table 2.1: Potential organisms used for the biosynthesis of nanoparticles

Nanomaterial	Size/nm	Agent	Type
Au	5–25	*Bacillus subtilis*	Bacteria
Au	10–20	*Shewanella algae*	Bacteria
Au, Ag, Au–Ag	20–50	*Lactobacillus* sp.	Bacteria
CdS	5–200	*Klebsiella pneumoniae*	Bacteria
Ag	6.4	*Aeromonas* sp. SH10	Bacteria
Au	10–50	*Penicillium brevicompactum*	Fungi
Au	20 ± 8	*Verticillium* sp.	Fungi
ZnO	1.2–6.8	*Aspergillus fumigatus*	Fungi
$FeCl_3$	10–24.6	*Aspergillus oryzae*	Fungi
$Ca_3P_2O_8$	28.2	*Aspergillus tubingensis*	Fungi
CdS	2-2.9	*Candida glabrata*	Yeast
CdS	1–1.5	*Schizosaccharomyces pombe*	Yeast
Ag	2-5	*Yeast strain* MKY3	Yeast

Source: Boroumand et al. 2015

Microorganism-mediated synthesis of nanoparticles happen in extracellular environment or intracellular environments and both of these processes involve several biochemical reactions, mostly mediated by enzymes. Generally, fungi, bacteria and plants are used in the process of synthesizing nanoparticles. Through the enzymes, it can play many functions such as changing the oxidation state of metals, controlling of size and shape of nanoparticles synthesizing, and aiding in reducing the size of metal particles synthesized into the nanoscale. However, the reaction rate of nanoparticles is largely dependent upon the microbial species involved, pH of the reaction medium, dissolved oxygen level, the metal concentration of the medium and their toxicity to cells and physical characteristics such as incubation temperature, agitation rate and exposure time.

Intracellular synthesis of nanomaterials involves three steps namely, trapping, bioreduction, and capping. Trapping involves catching metal ions by the cell surfaces with the aid of electrostatic interactions. Then it follows bioreduction which involves the reduction of metal ions to nanosize through enzyme-mediated reactions (Ahmad et al. 2002). Finally, capping involves the coating of various compounds such as surfactants around the particles enabling the nanoparticles to change their properties such as toxicity.

Fungi have well-defined micro-factories which release proteins and enzymes extracellularly (Ahmad et al. 2002, Syed and Ahmad 2012). Enzymes such as reductases released by them have been predominantly used. For instance, sulfate reductases and nitrate reductases are some enzyme categories which has been identified for green synthesis (Kumar et al. 2007, Gholami-Shabani et al. 2014). It was reported that proteins, flavonoids and triterpenoids released by fungi can reduce metallic ions to metallic nanoparticles. Also, stabilization of the nanoparticles is an important aspect of their synthesis and these compounds can act as capping and stabilizing agents during the synthesis. This is beneficial as it does not require distinct capping and stabilizing agents as in chemical-based methodologies. Free amino groups and carboxylate groups of proteins can reduce and stabilize metallic nanoparticles. Therefore, surface-bound proteins during the synthesis of nanomaterials using greener pathways can result in longer-term stability.

2.3.1 Potential for Green Synthesis of Nanoparticles and Advantages

- The low financial cost and scaling up potential due to low consumption of energy.
- Higher yield due to higher production rate
- Biological properties; least harmful to the environment and comparatively low toxicity to the organisms, higher solubility and biocompatibility.

- Physicochemical properties; higher stability, better crystallization and aggregation properties.

2.4 Conclusion and Future Perspectives

The synthesis of nanomaterials is the primary step in nanotechnology-based approaches. Imperatively maintaining the uniformity and desired properties is a basic requirement for their continuous application. Generally, biological, physical and chemical-based approaches are used to synthesize nanoparticles. However, every approach has its own merits and demerits. Currently, chemical-based strategies are widely being in application to their synthesis and most of the chemical byproducts and wastes generated can have toxic impacts on the environment. Therefore, nanotechnologists have to search for less toxic feasible greener approaches to nanomaterial synthesis. One such strategy is to synthesize nanoparticles using solvent-free methods. Also, microorganisms are considered micro-factories which release diverse types of reducing agents intracellularly and extracellularly during their metabolism. These metabolites can be harnessed to apply as potential agents in the green synthesis of nanomaterials. Although this has been identified as a better strategy it still requires thorough studies to enhance the quality of the nanoparticles synthesized.

References

Ahmad, A., P. Mukherjee, D. Mandal, S. Senapati, M.I. Khan, R. Kumar and M. Sastry. 2002. Enzyme mediated extracellular synthesis of CdS nanoparticles by the fungus, *Fusarium oxysporum*. *Journal of the American Chemical Society*, 124(41), pp. 12108–12109.

Baig, N., I. Kammakakam and W. Falath. 2021. Nanomaterials: A review of synthesis methods, properties, recent progress, and challenges. *Materials Advances*, 2(6), pp. 1821–1871.

Boroumand Moghaddam, A., F. Namvar, M. Moniri, S. Azizi and R. Mohamad. 2015. Nanoparticles biosynthesized by fungi and yeast: A review of their preparation, properties, and medical applications. *Molecules*, 20(9), pp. 16540–16565.

Carlsson, J.O. and P.M. Martin. 2010. Chemical vapor deposition. In: *Handbook of Deposition Technologies for Films and Coatings* (pp. 314–363). William Andrew Publishing.

Gholami-Shabani, M., A. Akbarzadeh, D. Norouzian, A. Amini, Z. Gholami-Shabani, A. Imani, M. Chiani, G. Riazi, M. Shams-Ghahfarokhi and M. Razzaghi-Abyaneh. 2014. Antimicrobial activity and physical characterization of silver nanoparticles green synthesized using nitrate reductase from *Fusarium oxysporum*. *Applied Biochemistry and Biotechnology*, 172(8), pp. 4084–4098.

Ijaz, I., E. Gilani, A. Nazir and A. Bukhari. 2020. Detail review on chemical, physical and green synthesis, classification, characterizations and applications of nanoparticles. *Green Chemistry Letters and Reviews*, 13(3), pp. 223–245.

Kolahalam, L.A., I.K. Viswanath, B.S. Diwakar, B. Govindh, V. Reddy and Y.L.N. Murthy. 2019. Review on nanomaterials: Synthesis and applications. *Materials Today: Proceedings*, 18, pp. 2182–2190.

Kumar, S.A., M.K. Abyaneh, S.W. Gosavi, S.K. Kulkarni, R. Pasricha, A. Ahmad and M.I. Khan. 2007. Nitrate reductase-mediated synthesis of silver nanoparticles from $AgNO_3$. *Biotechnology Letters*, 29(3), pp. 439–445.

Madanayake, N.H., R. Rienzie and N.M. Adassooriya. 2019. Nanoparticles in nanotheranostics applications. In: *Nanotheranostics* (pp. 19–40). Springer, Cham.

Narayanan, K.B. and N. Sakthivel. 2010. Biological synthesis of metal nanoparticles by microbes. *Advances in Colloid and Interface Science*, 156(1-2), pp. 1–13.

Pal, G., P. Rai and A. Pandey. 2019. Green synthesis of nanoparticles: A greener approach for a cleaner future. In: *Green Synthesis, Characterization and Applications of Nanoparticles* (pp. 1–26). Elsevier.

Pimpin, A. and W. Srituravanich. 2012. Review on micro- and nanolithography techniques and their applications. *Engineering Journal*, 16(1), pp. 37–56.

Rane, A.V., K. Kanny, V.K. Abitha and S. Thomas. 2018. Methods for synthesis of nanoparticles and fabrication of nanocomposites. In: *Synthesis of Inorganic Nanomaterials* (pp. 121–139). Woodhead Publishing.

Seetharaman, P.K., R. Chandrasekaran, S. Gnanasekar, G. Chandrakasan, M. Gupta, D.B. Manikandan and S. Sivaperumal. 2018. Antimicrobial and larvicidal activity of eco-friendly silver nanoparticles synthesized from endophytic fungi *Phomopsis liquidambaris*. *Biocatalysis and Agricultural Biotechnology*, 16, pp. 22–30.

Singh, V. and S.K. Chakarvarti. 2016. Biotemplates and their uses in nanomaterials synthesis: A review. *American Journal of Bioengineering and Biotechnology*, 2(1), pp. 1–14.

Sundaram, P.A., R. Augustine and M. Kannan. 2012. Extracellular biosynthesis of iron oxide nanoparticles by *Bacillus subtilis* strains isolated from rhizosphere soil. *Biotechnology and Bioprocess Engineering*, 17(4), pp. 835–840.

Syed, A. and A. Ahmad. 2012. Extracellular biosynthesis of platinum nanoparticles using the fungus *Fusarium oxysporum*. *Colloids and Surfaces B: Biointerfaces*, 97, pp. 27–31.

Nanomaterials Intake and Translocation in Plants

Nanomaterials are widely utilized in a variety of products leading to a significant increment in their production and use. The progressive increase in applications of nanomaterial-based products has become a huge concern due to its potential impact both on the flora and fauna in the environment. The higher reactivity of nanomaterials facilitates the direct interaction with both biotic and abiotic components. Currently, scientists have focused a lot on studying the possible impacts of nanomaterials on humans and the environment. Some of the engineered nanomaterials show the potential to become hazardous pollutants and sometimes they are regarded as emerging pollutants in the environment. However, the lack of available data on the actual release of nanomaterials to the environment limits the knowledge of their potential risks.

Nanomaterials are ubiquitous and they can be released into different environs via natural and artificial processes. These materials may finally end up in the air, water, soil or any other environmental matrix. Nanomaterials released into the environment from anthropogenic activities are a major concern among scientists to understand their fate and behavior. Most of them will finally end up in the soil which can be the main endpoint where nanomaterials will end up.

Plants are the pioneers in the biosphere which can self-synthesize their food by converting solar energy to chemical energy. Plants can directly acquire nanomaterials or nanoparticles available in the environment. The root system of the plants is the lifeline which accumulates water and mineral nutrients from the soil. They also, provide the strength to anchor the plant in the soil. Mechanisms involved in the uptake of plant nutrients also can take in nanomaterials through active and passive processes. Therefore, direct interaction of nanomaterials with plants can impose positive as well as negative impacts on plants.

3.1 The Release of Engineered Nanomaterials into the Environment

Life cycle assessments of nanomaterials will provide precise data on their release scenarios. Primarily nanomaterials are released into the environment both from direct and indirect pathways. They can be released into the environment during the synthesis, product manufacturing or directly in agricultural or remediation strategies. In addition, direct release is possible via open windows when powdery materials are used inattentively or from transport accidents or spills in industries.

Indirectly, release can occur from production sites into rivers or other water streams either via untreated or treated wastewater. Studies have shown that nanomaterials could be released into different environmental matrices arising from the production processes. It has been estimated that 0 to 2% of the produced nanomaterials reach the environment from the production procedure. However, release into the environment is process-dependent and proper equipment maintenance can mitigate their release to the environment significantly.

Nanomaterials which get released into the air are likely to be deposited in the environment sooner or later. In addition, sewage treatment plants or wastewater treatment plants can be identified as potential endpoints where nanomaterials get deposited. Much of the unintentional release will be into wastewater or solid waste. Wastewater treatment plants and waste incineration plants will be important sources of release. The main possibility for nanoparticles to be uncontrollably released into the environment is during the use, recycling and disposal of nanomaterial-containing products. The release of nanomaterials during such life stages of the nanomaterials is caused by the intended release from product applications.

3.2 Fate and Behavior of Nanomaterials in the Environment

Understanding the fate and behavior of nanomaterials in the environment will make a better foundation for the readers to understand the impact of nanomaterials. The environmental fate and behavior of nanomaterials thus released are fundamental to understand because their effects on environmental health are vital for risk assessment. In the previous section of this chapter, we have highlighted how nanomaterials will be released into the environment. Nanomaterials thus released can end up in different sinks. Most of these nanoparticles will end up in the soil. In addition, a certain proportion will be retained in water and air as well. However, exploring the behavior of such nanomaterials in the environment will allow us to assess specific hazards and potential impacts on each tropic level.

3.3 Physicochemical Properties Influencing the Fate and Behavior of Nanomaterials

Physicochemical properties of nanomaterials including size, shape, composition and surface energy influence their fate and behavior. This is important for two main reasons, firstly, for risk assessment and management purposes and to assess their pros and cons. Secondly, behaviors and their extent of effects are critically dependent on these properties. Intrinsic and extrinsic factors can have a significant impact on these behaviors. Nanomaterials transform within the environment and this can be significantly affected by their fate.

3.4 Processes and Transformations Affecting Nanomaterial Fate and Behavior

Transformations of nanomaterials result in changes in fate, behavior, and bioavailability, and have different effects on different environmental matrices. Dramatic improvements in our knowledge of transformations in complex media allow us to understand the effects on flora and fauna. Transformations of nanomaterials can be either physical, chemical or biological. Physical transformations include aggregation, agglomeration, sedimentation, and deposition. Chemical processes include dissolution and subsequent speciation changes, redox reactions (oxidation and sulfidation), photochemical reactions, and corona formation. Biologically mediated processes include biodegradation and biomodification which are mediated by microbes and exudates from other organisms. It is noteworthy to mention that these transformations are dependent on nanomaterial nature and the environmental conditions as mentioned previously.

3.4.1 Dissolution and Solubility

Dissolution of nanomaterials to release ions shows a direct impact on biological systems. Hence, nanomaterials enhance both transport and local ionic concentrations in a biological system. Nanomaterials such as zinc oxide can impose toxic effects when they undergo dissolution to release ions (Adele et al. 2018). Here the solubility has a major influence on the fate and toxicity of ZnO. In addition, nanomaterials including silver, copper oxide, and some quantum dots show intermediate solubilities and dissolution potentials. These forms of nanomaterials show their toxicological impacts depending on both ions and particles in a given system. Carbon-based nanomaterials, ceria and titania show a lower solubility compared to other nanomaterials. Therefore, dissolution and solubility of these nanomaterials become less important to show their impacts. Although different nanomaterials show variable dissolution potentials, their solubility is pH-dependent (Rathnayake et al. 2014). Hence, the conditions of the microenvironment where nanomaterials exist are utterly important to understand its fate and behavior. For example, hydroxyapatite nanoparticles easily undergo dissolution in acidic soil allowing plants to absorb them (Montalvo et al. 2015). In contrast, the same nanomaterial will tend to be found as precipitates or sediments when soil is alkaline. Also, the dissolution of nanomaterials can be judged over the exposure period in a given environment and this will alter their equilibrium solubility. Natural organic matters also show a positive impact on nanomaterial dissolution. Certain compounds can enhance this along with particle maturation and precipitation to form new materials at the nanoscale. Copper oxide nanomaterials undergo dissolution when they interact with soil organic matter (Hortin et al. 2020). This also highlights the role of soil organic matter to control the solubility of nanomaterials in the environment.

When a nanoparticulate matter enters a biological system, it reacts with cellular components as intact or as different forms derived from the main unit nanomaterial. For instance, in the case of metallic nanoparticles, the release of elemental form can occur as ions. If it is Ag nanoparticles releasing form will be Ag^+, if it is ZnO nanoparticles it often releases as Zn^{2+}. Certain nanoparticles further change their oxidation state. Then they can form complex structures with the dissolved compounds present in the cellular environment. In addition to subjecting to changes in oxidation state and taking part in chemical reactions, nanomaterials can accumulate within the plant organs such as fruits, tubers, leaves etc. causing different impacts on plant physiology.

State changes can also occur in the air too. For instance, the reaction with carbonic acid produced by dissolving CO_2 gas in water can react with nanomaterials present in the air and could convert them to ionic form. This reaction can also happen on plant surfaces too.

3.4.2 Aggregation, Sedimentation, and Deposition

The nature of the nanomaterials in different sinks and receptors may also be fundamentally different because of their aggregation and dispersion. Aggregation often reduces the number of bioavailable nanomaterials in the environment. Still, there is no evidence of the bioavailability of nanomaterials once they are ingested into biological systems. Sometimes, aggregation can improve the bioaccumulation of nanomaterials in different tropics.

Aggregation occurs due to the sum of the interaction forces between particles. When attractive forces of nanomaterials prevail over the repulsive forces, particles tend to aggregate in clusters and vice versa. The interactions between charged particles in a colloidal system in a solution are described by the DLVO theory. The DLVO theory (named after Boris Derjaguin and Lev Landau, Evert Verwey and Theodoor **O**verbeek) explains the aggregation of particles in aqueous dispersions as a combination of repulsive and attractive forces. This theory works well to emphasize charge stabilization in colloidal systems. The aggregation of nanomaterials depends on pH, ionic strength, the presence of ions, and the type and content of organic matter. Also, nanoparticle concentration will play a key role in aggregation. In general, nanomaterials can undergo homo-aggregation to produce clusters containing particles or colloids over 100 nm in size. However, it was reported that natural organic matters such as humic acid and fulvic acids can prevent nanomaterials from aggregation by stabilizing them via electrostatic and steric stabilization. Furthermore, aggregation between non-similar particles can take place leading to hetero-aggregation at higher doses of nanoparticles.

In addition, surface stabilization oversights the aggregation potential. For instance, silver nanoparticles show slow aggregation in low ionic strength solutions once it is stabilized with negative charges on the surface. This shows the electrostatic stabilization of nanoparticles where zeta potential is used for colloidal stability. Surface coating such as polyvinyl pyrrolidone can stabilize nanoparticles by steric stabilization. Therefore, coating nanoparticles with natural organic matter can stabilize them by steric stabilization due to steric repulsion. Sometimes, the opposite can happen in the presence of certain natural organic matter. Hence, these can control the stabilization or destabilization of environmental systems.

3.4.3 Sulphidation and Redox Behavior

Sulphidation is a major chemical transformation for many metallic nanoparticles. This phenomenon occurs in the presence of higher doses of sulfide sources and can be found in wastewater or anoxic sediments. Sulphidation can result in clear changes to particle size, surface charge, and solubility. For the sulphidation to occur it requires both oxygen and sulfide to directly react at the surface of the nanomaterial. Indirectly, this can take place by releasing metallic ions followed by precipitation as the metal sulphide. Oxy-sulphidation is one of the most preferred routes of nanomaterial modification when sulphide concentrations are higher in the environment. However, the presence of natural organic matter plays a protective role in reducing sulphidation rates. Sulphidation of metallic nanomaterials can be controlled by controlling the dissolution rates. Metallic nanomaterials such as silver, iron, copper and zinc are susceptible to sulphidation, however, their rates of sulphidation can vary on the physicochemical properties of nanomaterials and the environmental conditions.

3.5 Nanomaterials Uptake, Translocation and Deposition by Plants

Nanomaterials can interact easily with plants directly or indirectly. Directly they can interact with plants once they are utilized in agricultural applications. Also, indirectly they interact with plants when they are released during other applications. Plants are a vital component of ecological systems and serve both as important ecological receptors and as a potential route for the transportation and bioaccumulation of nanomaterials at different trophic levels.

The first foremost candidate for nanomaterials to interact with, will be plants. Hence, the uptake of nanomaterials into plants can be vital because the plant can be considered the pioneer at every trophic level. Therefore, understanding the factors involved in nanomaterial uptake will be imperative. The main limitation of the nanomaterial penetration into the plants is the size. In addition, nanomaterial type and its chemical composition also influence. Furthermore, the morphology, functionalization and coating of the nanomaterial surface can greatly alter the properties of its absorption and accumulation by the plant.

Size is a critical factor affecting nanoparticle absorption by plants. The route of nanomaterial entry depends on the size exclusion limit. For instance, the size exclusion limit for polar and ionic solutes to penetrate the cuticle membrane of plants is ranged from 2~2.4 nm. In addition, the tiny openings at the leaf surfaces known as stomata are another opening for the entry of nanomaterials into the plants. Generally, plant leaves consist of 100–1300 stomata per mm^2 of the leaf surface. Stomata is a cellular pore formed between two guard cells. Dimensionally, it has an average width of 3–12 μm and a length of 30 μm for gaseous exchange. It has been shown that nanomaterials of sizes above 20 nm can diffuse via a stomatal pathway. Therefore, it is a vital parameter that warrants its entrance through cell wall pores or plant stomata and other entry points. During foliar application, nanomaterials tend to aggregate at the surface of the leaves. Hence, the uptake of aggregated nanomaterials is inhibited by the cuticular uptake. However, the stomatal pathway can act as the preferred pathway for such nanomaterials.

Studies have shown that ceria nanoparticles of particle size 7 nm were able to enter the root system and translocate within the plant. However, ceria nanoparticles with a particle size of 25 nm failed to cross the plant root system into cucumber plants where they were only detected on the outer surface of the cucumber roots. Another study reported that gold nanoparticles with a diameter of 3.5 nm entered tobacco root systems while larger (18 nm) particles were unable to get into the root epidermis. These observations confirm the size of nanomaterials directly influences plant uptake.

Surface charge is an important property which determines the interaction potential of nanomaterials with plants. The interaction of charged biological structures can determine the ability of nanomaterials to cross biological structures. Furthermore, charge affects the spatial distribution of nanoparticles that have translocated within plants. The translocation of nanoparticles can be affected by the surface charge. It was observed that negatively charged nanomaterials could translocate more efficiently than positively charged particles. Positively charged nanoparticles adhere more strongly to the root surface limiting its translocation. Therefore, negatively charged nanoparticles will efficiently translocate than positively charged nanoparticles in plants. Neutral particles show an intermediate degree of interaction concerning their translocation efficiency (Liu et al. 2019, Spielman-Sun et al. 2019).

Uptake and accumulation of nanomaterials by plants depend on the exposure concentrations as well. However, the movement and their uptake through the plants seem to be low and it is slower when it is in soil. Naturally forming pores during the biological processes, due to root exudates, fungal hyphae, and bacteria cells considerably modify the soil porous network and transportation of nanomaterials.

The shape also plays an important role in the uptake of nanomaterials by plants. The shape of nanomaterials relates to their aspect ratio. For instance, smaller droplet size aerosols containing nanomaterials enhance the translocation due to low aspect ratio and vice versa.

The uptake of nanoparticles by plants can be boosted in the presence of surfactants. Surfactants enhance the permissibility of the cuticle for nanoparticles. Certain studies have reported that the addition of a surfactant permits the accumulation of quantum dots in maize. Fragile porous regions can be observed in cuticles. For example, trichrome, root hairs and tips covered with mucilage and hydathodes can enhance the uptake of nanomaterials with fewer restrictions. Aquatic plants lack cuticular waxes on the leaves and their cuticle is thinner compared to terrestrial plants. This enables them to absorb nutrients and accumulate them in the leaves. Therefore, these plants can uptake nanoparticles without any difficulty in larger quantities compared to other plants (Banerjee et al. 2019).

Microorganisms in plants produce various symbiotic microenvironments around them. Symbiotic microenvironments can influence the metal and oxides of metallic nanoparticles' movement and uptake in plants. For example, nitrogen-fixing species such as *Rhizobium leguminosarum* establish symbiotic relationships with leguminous plants. During the nodulation, it generates pores of a few micrometers in size at root hair. These newly-formed nodules can act as hotspots for the uptake of heavy metals and some nonmetals (e.g., P, Se) or nanoparticles. Also, vesicular mycorrhizal fungi form symbiotic relationships with more than 80% of plants. This is a vital factor to provide the plants with nutrients and water to increase the host's resistance to biotic and abiotic stresses. This can directly influence the plant's health and productivity. Therefore, mycorrhiza fungi can influence nanoparticle uptake by plants thus directly influencing plants (Judy et al. 2016, Noori et al. 2017).

3.6 Uptake and Translocation of Nanomaterials

In recent times, much advancement has been taken in knowing the mechanisms of uptake and translocation to trace the kinetics of engineered nanoparticles. Different nanomaterials can enter the vegetative systems either via foliar or root pathways. Foliar uptake of nanomaterials involves the entry of nanomaterials into the plants through natural openings present in the shoots of plants. Figure 3.1 depicts the possible modes of nanomaterial entry from the aerial parts of a plant. Also, they can evade the plant systems through cracks and wounds as well. Primarily, leaves are defined as a potential route for the uptake of nanoparticles. The cuticle layer at the leaf surface is considered as the first barrier to the foliar uptake of nanomaterials. The cuticle layer of the plant gives a defensive mechanism to the whole system from excessive water loss, and competently protects the plant from the attack of pathogens, foreign particles, and dirt. Therefore, the cuticle is found to be the first obstruction for nanoparticle uptake in all sections of the plant. Cuticles are heterogeneous membranes in the plant leaf surfaces. The positional variations in the composition and the solute properties of the cuticle determine their uptake. Furthermore, the permeability of nanomaterials has an inverse relationship with the thickness of the membrane. In other words, an increase in diffusion path length can decrease its uptake to the plant through leaves. Also, the permeability of nanomaterials can be enhanced with the increase of solubility in cutin and amorphous

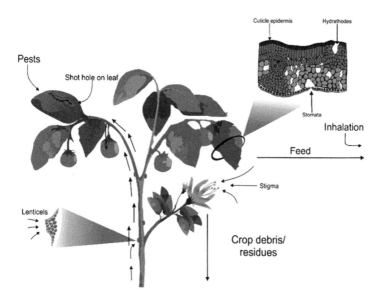

Figure 3.1: Entry of nanomaterials from aerial parts and top-down movement.

waxes of the leaf cuticle. After nanomaterials are contacted with roots and they penetrate the cell wall and then to the cell membrane through the pores of the epidermis. From the epidermis, they move to the cortex, cambium and vascular tissues and then to the stele region of the roots (see Figure 3.2). From there, the translocation occurs via the xylem of the plant (Raliya et al. 2016).

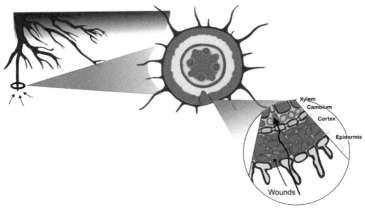

Figure 3.2: Entry of nanomaterials from roots and bottom-up movement.

Cellular pathways for the transport of nanoparticles are carried out by both apoplast and symplast pathways (see figure 3.3). The apoplast is the non-living environment outside the plasma membrane which includes the cell wall and intercellular spaces through which substances move freely. These substances cannot move freely all the time via this path, because of the Casparian strip of endodermal cells, and air spaces between plant cells and cuticles. Thus, they need to follow different means of transportation which are, for example, the symplast pathway which is the cytoplasm-mediated pathway that facilitates the movement of substances between cells via plasmodesmata. Studies have shown that the apoplast pathway prefers the transportation of larger particles (~200 nm) and the symplast pathway favors the translocation of smaller (<50 nm) particles. Furthermore, size-dependent cellular transport pathways explain that the apoplast pathway is predominant when nanomaterials are applied to a plant in the drop cast method. In contrast, symplast pathways are common for aerosol-mediated delivery of nanomaterials. It is also

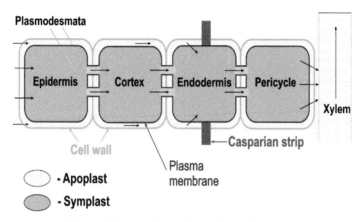

Figure 3.3: Apoplast and symplast pathways.

reported that the processes of nanoparticle uptake are active-transport mechanisms and it involves several additional cellular mechanisms, such as signalling, recycling, and the regulation of the plasma membrane. In addition, physicochemical fluctuations in various plants (variation in hydraulic conductivity, size of the pore of the cell wall) might influence on the passage and gathering of nanoparticles.

Nanomaterials form complexes with transporter proteins at the cell plasma membrane or from root exudates. This will mediate the entry of nanomaterials into the plants. Translocation of nanomaterials can take place from root to leaf/fruit or from leaf to roots. Once it enters the plants from roots, nanomaterials get diffused between plant cell walls and plasma membrane. Then the osmotic pressure and capillary forces within the plant influence its movements. In addition, transporter proteins such as aquaporin and ion channels may also involve in the entry of nanomaterials via endocytosis or piercing the cell membrane. The major protein involved in the transportation of nanomaterials is aquaporins of which the monomer size is around 30 kDa. Nanomaterials also can transport through the cell membrane with the aid of a group of proteins named ion channels. They provide the passage for ions across a membrane by forming pores. Here, the nanoparticles when contacted with the cellular membrane, it forms a bud-like structure surrounding the nanomaterials. This structure is formed by the cell membrane itself and consists of the lipid bilayer.

Endocytic uptake occurs when specific receptor-ligand interactions take place. Moreover, endocytosis can occur via clathrin-mediated and clathrin-independent pathways. Also, nanoparticles are transported through new pores across the cell membrane. Figure 3.4 elaborates the different mechanisms involved in the entry of nanoparticles into plant cells.

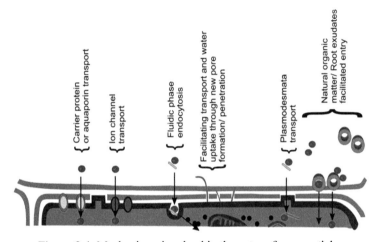

Figure 3.4: Mechanisms involved in the entry of nanoparticles into plant cells.

Once the nanomaterials are entered into the cytoplasm, they are further transported throughout the system through the apoplastic pathway, symplastic pathway and then by cell-to-cell movement using plasmodesmata. Entry to the stele from the vascular tissues (xylem) occurs via a symplastic pathway. Plasmodesmata are cytoplasmic interconnections present between two plant cells that facilitate the movement of substances (Banerjee et al. 2019, Ali et al. 2021).

Depending on the morphology of the engineered nanomaterial, they can enter the plants by piercing the cell membranes to enter the cell cytosol. Carbon-based nanomaterials reported showing such a mode of entry into plants. The high reactivity of nanomaterials allows them to interact with biomolecules inside the cytoplasm by van der Waals, electrostatic, and steric-polymer forces. These interactions can result in forming protein coronas. These internalized complexes can move to adjacent cells through plasmodesmata which are having diameters in the range of 20–50 nm. Plasmodesmata are microscopic channels that link the cell walls of adolescent plant cells. They provide the passage for nanoparticles to move from one cell to the other. However, the sustainability of the plasmodesmata can be affected by nanomaterials by disrupting the cytoskeleton microfilaments. Foliar application of nanomaterials allows them to penetrate leaves through stomatal openings. Also, cuticular pathways are involved in nanomaterials uptake as mentioned in previous sections. In addition, nanomaterials can penetrate cells from the bases of trichomes and are subsequently translocated through the vascular system (Ali et al. 2021).

3.7 Conclusion and Future Perspectives

Plants are the principal primary producers in most ecosystems. The deliberate and accidental release of nanomaterials into the ecosystem has become a huge concern due to its potential impacts both on flora and fauna due to the higher reactivity of nanomaterials with both biotic and abiotic components. Hence, studying the interaction of nanomaterials with plants would also help gain information on the potential impact created by them. Plants show positive as well as negative impacts based on the type and properties of nanoparticles. Also, the uptake and translocation of nanomaterials in plants would be imperative to understand. However, still there is huge demand and a gap in the studies in this area. Furthermore, the lack of available data on the actual release of nanomaterials to the environment limits the knowledge of how nanomaterials will affect different ecosystems.

References

Adele, N.C., B.T. Ngwenya, K.V. Heal and J.F.W. Mosselmans. 2018. Soil bacteria override speciation effects on zinc phytotoxicity in zinc-contaminated soils. *Environmental Science & Technology*, 52(6), pp. 3412–3421.

Ali, S., A. Mehmood and N. Khan. 2021. Uptake, translocation, and consequences of nanomaterials on plant growth and stress adaptation. *Journal of Nanomaterials*, 2021.

Banerjee, K., P. Pramanik, A. Maity, D.C. Joshi, S.H. Wani and P. Krishnan. 2019. Methods of using nanomaterials to plant systems and their delivery to plants (Mode of entry, uptake, translocation, accumulation, biotransformation and barriers). In: *Advances in Phytonanotechnology* (pp. 123–152). Academic Press.

Hortin, J.M., A.J. Anderson, D.W. Britt, A.R. Jacobson and J.E. McLean. 2020. Copper oxide nanoparticle dissolution at alkaline pH is controlled by dissolved organic matter: Influence of soil-derived organic matter, wheat, bacteria, and nanoparticle coating. *Environmental Science: Nano*, 7(9), pp. 2618–2631.

Judy, J.D., J.K. Kirby, M.J. McLaughlin, T. Cavagnaro and P.M. Bertsch. 2016. Gold nanomaterial uptake from soil is not increased by arbuscular mycorrhizal colonization of *Solanum Lycopersicum* (Tomato). *Nanomaterials*, 6(4), p. 68.

Liu, M., S. Feng, Y. Ma, C. Xie, X. He, Y. Ding, J. Zhang, W. Luo, L. Zheng, D. Chen and F. Yang. 2019. Influence of surface charge on the phytotoxicity, transformation, and translocation of CeO_2 nanoparticles in cucumber plants. *ACS Applied Materials & Interfaces*, 11(18), pp. 16905–16913.

Montalvo, D., M.J. McLaughlin and F. Degryse. 2015. Efficacy of hydroxyapatite nanoparticles as phosphorus fertilizer in andisols and oxisols. *Soil Science Society of America Journal*, 79(2), pp. 551–558.

Noori, A., J.C. White and L.A. Newman. 2017. Mycorrhizal fungi influence on silver uptake and membrane protein gene expression following silver nanoparticle exposure. *Journal of Nanoparticle Research*, 19(2), p. 66.

Raliya, R., C. Franke, S. Chavalmane, R. Nair, N. Reed and P. Biswas. 2016. Quantitative understanding of nanoparticle uptake in watermelon plants. *Frontiers in Plant Science*, 7, p. 1288.

Rathnayake, S., J.M. Unrine, J. Judy, A.F. Miller, W. Rao and P.M. Bertsch. 2014. Multitechnique investigation of the pH dependence of phosphate induced transformations of ZnO nanoparticles. *Environmental Science & Technology*, 48(9), pp. 4757–4764.

Spielman-Sun, E., A. Avellan, G.D. Bland, R.V. Tappero, A.S. Acerbo, J.M. Unrine, J.P. Giraldo and G.V. Lowry. 2019. Nanoparticle surface charge influences translocation and leaf distribution in vascular plants with contrasting anatomy. *Environmental Science: Nano*, 6(8), pp. 2508–2519.

Nanodelivery Systems of Agrochemicals for Plants

4.1 Controlled and Targeted Delivery of Agrochemicals

Plant nutrient formulations and pesticides that include insecticides and other chemicals are used to kill higher (rodents) and lower-order organisms (bacteria). Also, fungicides, herbicides (plant regulator, defoliant, or desiccant) and nitrogen stabilizers come under agrochemicals. These formulations consist of two major components namely active ingredients and inert ingredients. Active ingredients play a key role in the agrochemical process causing a direct impact on the target, while inert ingredients are the ones that contain additives and they aid in adhering to the surfaces, enhancing solubility and dispersal efficiency, protecting the active ingredients etc. Formulation of agrochemicals is essential and it involves processing its active ingredient to improve the storage, ease and safety of handling, the safety of the active ingredient and efficacy.

In order to achieve optimized growth and improved health status of crop plants, agrochemicals such as plant nutrients, pesticides and growth regulators are used. However, the efficiency of conventional approaches is low and it made scientists introduce novel alternatives to improve them for agricultural applications. Most conventional agrochemicals do not reach the target, thus leading to considerable losses. They are primarily attributed to their boost release and then accumulation in the ecosystems. Normally losses can be in the range of 60–70% in terms of pesticides, while in critical cases that can increase up to 95%. Conventional fertilizers can also be lost by about 40–75%. Therefore, developing nanotechnology-based strategies using nanoparticles for agricultural applications have identified as much more cost-effective and easier, because of the availability of an array of approved inert materials for use in agrochemicals.

Through nano-based delivery systems, they have been enabled to deliver the agrochemicals to the target in an efficient manner. Furthermore, depending on the nanomaterials composition used in the composites, it can alter the release time of a given target molecule or compound. Controlled release formulations of agrochemicals are now becoming popular. In addition to the agrochemicals used for crop production, certain drugs that are used for the control of farm animal diseases can also be incorporated with controlled release matrices. Therefore, another advantage created through nanotechnology is the ability to use lower quantities than those employed in conventional formulations (Roy et al. 2014). Figure 4.1 shows the release behavior of agrochemicals when in the forms of conventional and controlled release.

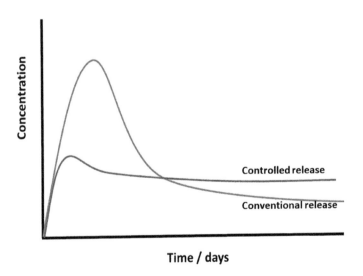

Figure 4.1: Behavior of release of active ingredients when in the form of conventional release and controlled release.

4.2 Advantages of Controlled-release Formulations over Conventional Formulations

The inclusion of different agrochemical formulations into nanomatrices has several advantages over the conventional approaches and these merits are elaborated below (Rahman et al. 2021).

1. Enhanced solubility of poorly soluble pesticides
   Many of the commercially available insecticide formulations are organic in chemical nature, hence they are poorly soluble in water (nearly 1% of aqueous saturation solubility). Therefore, organic solvents are used to dissolve these active ingredients, while if the resultant solution is hydrophobic, it can be highly prone to localized precipitation in the fields.

2. Enhanced efficacy of agrochemicals
   Efficacy can be simply defined overall as the pesticide's ability to produce desired results (a measure of its overall negative and positive effects). Here, it specifically matters in the ability of the pesticide to make contact with the target i.e. adhesion to the live surfaces and penetration through. In conventional fertilizer formulations, one of the unique problems is low use efficiency and low uptake potential by the plants. Through nano-based controlled release formulations it can overcome this inefficacious nature. For instance, urea-functionalized hydroxyapatite nanoparticles, amorphous calcium phosphate nano fertilizers and Fe_3O_4-urea nanocomposites are some of the recently reported nanoformulations for controlled and sustained release of nitrogen (Kottegoda et al. 2011, Carmona et al. 2021, Gaiotti et al. 2021, Guha et al. 2022).

3. Due to the targeted activity of certain controlled-release formulations and extended effective periods, it can reduce the frequency of application.
   Conventional strategies make different agrochemicals quickly released and accumulate in the environment reducing their chemical use efficiency by plants. However, controlled and sustained release using nanotechnology-based approaches can release these active ingredients slowly for extended

durations. Hence, controlled and slow-release nanosystems can significantly affect crop production and nanoformulations thus developed can regulate the nutrient supply with time, increase nutrient use efficiency and reduce environmental pollution. More importantly, extended durations of fertilizers and active target release can alter the frequency of application and is another benefit of using nanodelivery systems.

4. Reduced toxicity to non-targeted organisms
   Controlled release formulations can reduce toxicity effects on beneficial organisms such as earthworms, birds, and natural enemies.

5. Reduced premature degradation losses due to environmental conditions
   Environmental factors namely, chemical (chemical reactions with other compounds present in the soil, adsorption to soil particles), physical (sunlight), and biological (macro and microorganisms).

6. Reduced risk of contamination of other environmental systems such as ground and surface water systems
   Conventional agrochemical formulations are more prone to localized precipitation when in the presence of water in the microenvironment.

7. Enhanced safety for users and applicators due to reduced dosage per application and reduced treatment frequency.

8. Cost-effective nature of pest control.

4.2.1 Types of Potential Nanocarrier Systems that can be Used in Agricultural Applications, their Advantages and Disadvantages

Various nanomaterials have been used as nanocarriers to hold and release various agrochemicals in a controlled and sustained manner. In this regard, mesoporous silicon-based materials, solid lipidic nanoparticles (SLNs), nano-emulsions, dendrimers, nanocrystals, hydrogels and nanoclays have been widely researched.

Mesoporous Silicon-based Materials

These materials have a stable structure with tunable and controlled release properties. However, they are inorganic in nature, non-biodegradable and ability to cause cellular toxicity.

Solid Lipidic Nanoparticles (SLNs)

SLNs are biocompatible and biodegradable. Also, they can improve the solubility of poorly soluble active ingredients. However, they have low loading capacity and encapsulation efficiency while with the potential of releasing the active ingredient during storage.

Nano-emulsions

Nano-emulsions are highly stable for gravitational separation and aggregation. They can improve the solubility of active ingredients. Thus, compatible with lipophilic active ingredient molecules. They can also improve the efficacy of antimicrobial agents. Another advantage is their biocompatibility and biodegradability.

Dendrimers

Dendrimers are suited to be incorporated with lipophilic or lipophobic active ingredients and some dendrimer species are reported to be resistant to hydrolysis. Conversely, the toxicity is correlated with the number of NH_2 groups and thus reported as cytotoxic.

Nanocrystals

Nanocrystals can improve the bioavailability of water-insoluble compounds, increases drug adhesiveness to surface cell membranes, and enhances particle stability in suspension. However, the morphology and crystallinity of the final product are difficult to be controlled.

Hydrogels

Hydrogels are highly environmentally friendly. However, their batch-wise variation due to heterogeneity hinders performance.

Nanoclays

Nanoclays such as montmorillonite (MMT), bentonite, Kaolinite, and halloysite have been widely researched as matrices and these clay matrices have been treated with other chemical compounds such as acids, bases (strong or weak), and surfactants to enhance their efficiency. Table 4.1 will provide an overview of research evidence

Table 4.1: Agrochemical-nanoclay matrices

Purpose	Active ingredient	Chemical category	Clay type	Formulation	T50 (d)/desorption rate in water	References
CR	Amitrole	H	Kaolinite, Halloysite	(a) Amitrole-Kaolinite (b) Amitrole-Halloysite	(a) 50–80% (5 h), (b) 85–90% (5 h)	Tan et al. 2015
CR	Isoproturon	H	Bentonite	Isoproturon-Alginate-Bentonite	5.88–27.41 d (T50)	Garrido-Herrera et al. 2006
CR	Hexazinone	H	MMT	Hexazinone-MMT modified with Fe	0.01–88 d (T50)	Celis et al. 2002
Ads	Thiram	Fc	Bentonite	Thiram-Neem leaf powder-Alginate-Bentonite	2.5–3.5 mg/g (48 h)	Singh et al. 2010
CR	Urea	F	MMT	MMT–Urea	Pure urea; 100% and the composite, 16-56% (24 h)	Yamamoto et al. 2016
CR	Phosphate	F	Bentonite	Chitosan-Bentonite Beads-chitosan	30–99%	Piluharto et al. 2017

Ads: Adsorbant, CR: Controlled release, H: Herbicide, I: Insecticide, Fc: Fungicide, F: Fertilizer, T50: Time is taken to release 50% of the active ingredient.

on the controlled release of agrochemicals incorporated with nano clay matrices.

4.3 Enhancers of Soil Water Retention

Conservation of water is a major concern in present-day agriculture. Likewise, soil water distribution, storage, and saving have been enhanced by various nanotechnological methods and such related products are available in the world market. Certain materials such as polymers are widely used together with other nanomaterials to achieve these goals. Polymers and nanoclays are widely used in this context. Certain nanomaterial-based products like nanohydrogels have the proven ability to increase the water-holding capacity of soils. This property holds great promise for those dry areas where crop production is hindered due to water shortage. Moreover, sensors can be designed to detect water deficiency and plant stresses caused by it (Pulimi and Subramanian 2016).

4.4 Targeted Drug Delivery in Humans and Animals

When a drug enters a body of a living being, it has to reach up to the target cells which are affected. There are specialized molecules called receptors on cell surfaces. Drugs bind to these receptors and enter the cells. However, there are instances in that drugs do not reach their targets. One reason behind not delivering drugs to the desired site is their poor solubility. Nanotechnology has provided a greater opportunity in overcoming this problem. Several nanoparticles have proven to be effective in achieving this objective. Figure 4.2 explains the mechanism of targeted drug delivery.

4.4.1 Nanomaterials in Gene Delivery

Genetically modified crops include crops with higher resistance to herbicides, insects, diseases, and other biotic and abiotic stresses that have been generated using genetic engineering. Also, genetic manipulations in commercial crops enable to enhance the crop productivity and nutritional quality to fulfil the increasing global food

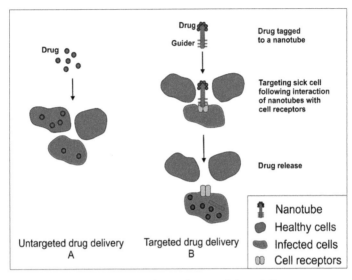

Figure 4.2: Mechanisms of targeted drug delivery.

demand. However, gene delivery into target host cells is a long and tedious task. For instance, *Agrobacterium*-mediated gene transfer is one such conventional approach which has been used for genetically modified crops (Lacroix and Citovsky 2019). Preparing genetically modified crops requires long-term tissue culture practices to recover transgenic plants and also it is limited by a narrow range of characters that can be transformed within a given crop species. In addition, limitations in the host range of different *Agrobacterium* species make them insufficient to address the increasing global demand. The intrusion of nanotechnology in drug delivery applications has allowed scientists to utilize it as a gene delivery platform. Techniques like magnetofection make the gene transfer to plant pollens more easily than conventional approaches thus facilitating a novel strategy in gene delivery to plants (Gad et al. 2020). Figure 4.3 shows how nanocarries can be used as gene delivery systems.

4.4.2　Evaluation of Saturation Solubility

Agrochemical or drug release profiles are essential to know how the mass transport mechanisms work concerning a given formulation.

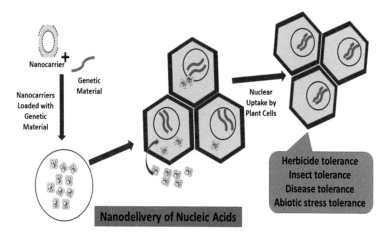

Figure 4.3: Nanocarriers for gene delivery.

Several mathematical relationships have been formulated by scientists to explain the release behavior of agrochemicals and how the release behavior impacts the efficiency of the application. It is very important to know how to use these equations to understand the different factors that affect the release velocity and how the dissolution behaviors can vary and influence the efficiency of the therapeutic regimen for patients.

Oswald-Freundlich Equation

According to the Oswald-Freundlich equation, the reduction of the particle size from the bulk level to the nanoscale enhances the saturation solubility of the material. Accordingly, this equation provides a way to predict an increase in saturation solubility with particle size reduction.

$$\frac{S(d)}{S(0)} = \exp \frac{\gamma Vm}{RTD}$$

In the above equation, $S(d)$ denotes the solubility of particles of which the diameter is D at temperature T. Vm is the molar volume and γ is the surface energy. R stands for the gas constant, and $S(0)$ is the solubility of the bulk material (Margulis-Goshen and Magdassi 2013).

Noyes-Whitney Equation

According to the above equation, the dissolution rate of a solid particle in aqueous media is inversely proportional to the particle radius, hence the dissolution rate increase with the reduced particle size.

$$\frac{dW}{dt} = \frac{DA(Cs - C)}{L}$$

Where dW/dt is the rate of dissolution, D represents the diffusion coefficient of the solid, L is the effective diffusion layer thickness, A is the surface area of the solid, Cs is the concentration of the solid in the diffusion layer surrounding the solid, and C is the concentration of the solid in dissolution medium. Upon a significant reduction of the particle radius achieved by nanosizing, A, the surface area, increases extremely, resulting in a much faster dissolution. Furthermore, enhancement of the solubility is achieved through the generation methods that result in partially or fully amorphous products (Margulis-Goshen and Magdassi 2013).

4.4.3 Evaluation of Release of Components from a Chemical Delivery System

The release of a chemical i.e., a drug or pesticide is an important characteristic of a chemical delivery system (Vega-Vásquez et al. 2020). A scientist named A. Fick developed laws of diffusion which are inevitable when studying the release of such components.

Fick's First Law

The diffusion flux due to diffusion is directly proportional to the concentration gradient.

$$J \propto \Delta C_A / \Delta X$$

Fick's Second Law

Fick's second law is an extension of the first law. It relates together to the change of concentration, change in time, and change in position. Therefore, the second law is time-dependent.

$$\Delta C_A / \Delta t = \Delta / \Delta X. \, [D_{AB} \cdot \Delta C_A / \Delta X]$$

Therefore,

$$\Delta C_A / \Delta t = D_{AB} . \Delta^2 C_A / \Delta X^2$$

Evaluation of nanopesticides or any pesticide can be performed using various mathematical models and some of which are frequently used are mentioned below (Mircioiu et al. 2019).

First-order Model

$$\frac{dc}{dt} = -Kc$$

$$\log C = \log C_0 - K_t / 2.303$$

Where C_0 represents initial concentration and C is the released concentration of bioactive molecules at time t, and K_t is the rate constant.

Higuchi Model

To determine the release kinetics of bioactive molecules from semi-solid and solid polymeric devices.

$$Q = K_H \, t^{1/2}$$

Where Q is the release concentration of bioactive molecules at time t, and K_H is the Higuchi dissolution constant.

Ritger-Peppas and Korsmeyer-Peppas Models

Evaluation of the release kinetics of bioactive molecules from swellable or non-swellable polymeric release systems, Ritger-Peppas and Korsmeyer-Peppas models can be employed as mathematical models to analyze Fickian and non-Fickian release.

Ritger-Peppas Model

$$W_t / W_\infty = k.t^n$$

Korsmeyer-Peppas Model

$$W_t / W_\infty = (D.t)^{0.5} / \pi L^2$$

Where in both models, W_t and W_∞ represent concentrations of bioactive molecules released at t time and at equilibrium, respectively. In the

Ritger-Peppas model, n is the release exponent, k is the swelling factor while in the Korsmeyer-Peppas model D is the diffusion constant, and L is the diameter of the dry nanoparticles.

4.5 Conclusion and Future Perspectives

Controlled delivery of agrochemicals using nanomatrices has been gaining attention to minimize the negative impacts of agricultural formulations. Moreover, drug formulations with controlled and sustained release behavior, targeting humans and animals have also gained greater attention. There is no argument that these formulations minimize the cost while achieving greater efficacy in terms of the desired effect. Besides, the many absorption matrices have also been formulated and tested to enhance soil water retention and water purification. Mathematical modelling is very popular and highly used to study the release kinetics of such formulations hence, providing greater value for the formulations. Indeed, the above matrices have shown promising results at the laboratory scale in most instances, however, still, the information related to economics and scalability is lacking. Therefore, pilot scale commercialization or scale-up projects are needed to reap the maximum out of the research findings. Hence future research should be based partly on feasibility and pilot scale-up projects.

References

Carmona, F.J., G. Dal Sasso, G.B. Ramírez-Rodríguez, Y. Pii, J.M. Delgado-López, A. Guagliardi and N. Masciocchi. 2021. Urea-functionalized amorphous calcium phosphate nanofertilizers: Optimizing the synthetic strategy towards environmental sustainability and manufacturing costs. *Scientific Reports*, 11(1), pp. 1–14.

Celis, R., M.C. Hermosín, M.J. Carrizosa and J. Cornejo. 2002. Inorganic and organic clays as carriers for controlled release of the herbicide hexazinone. *Journal of Agricultural and Food Chemistry*, 50(8), pp. 2324–2330.

Gad, M.A., M.J. Li, F.K. Ahmed and H. Almoammar. 2020. Nanomaterials for gene delivery and editing in plants: Challenges and future perspective.

In: *Multifunctional Hybrid Nanomaterials for Sustainable Agri-Food and Ecosystems* (pp. 135–153). Elsevier.

Gaiotti, F., M. Lucchetta, G. Rodegher, D. Lorenzoni, E. Longo, E. Boselli, S. Cesco, N. Belfiore, L. Lovat, J.M. Delgado-López and F.J. Carmona. 2021. Urea-doped calcium phosphate nanoparticles as sustainable nitrogen nanofertilizers for viticulture: Implications on yield and quality of Pinot gris grapevines. *Agronomy*, 11(6), 1026.

Garrido-Herrera, F.J., E. Gonzalez-Pradas and M. Fernández-Pérez. 2006. Controlled release of isoproturon, imidacloprid, and cyromazine from alginate-bentonite-activated carbon formulations. *Journal of Agricultural and Food Chemistry*, 54(26), pp. 10053–10060.

Guha, T., G. Gopal, A. Mukherjee and R. Kundu. 2022. Fe_3O_4-urea nanocomposites as a novel nitrogen fertilizer for improving nutrient utilization efficiency and reducing environmental pollution. *Environmental Pollution*, 292, 118301.

Kottegoda, N., I. Munaweera, N. Madusanka and V. Karunaratne. 2011. A green slow-release fertilizer composition based on urea-modified hydroxyapatite nanoparticles encapsulated wood. *Current Science*, pp. 73–78.

Lacroix, B. and V. Citovsky. 2019. Pathways of DNA transfer to plants from Agrobacterium tumefaciens and related bacterial species. *Annual Review of Phytopathology*, 57, p. 231.

Margulis-Goshen, K. and S. Magdassi. 2013. Nanotechnology: An advanced approach to the development of potent insecticides. In: *Advanced Technologies for Managing Insect Pests* (pp. 295–314). Springer, Dordrecht.

Mircioiu, C., V. Voicu, V. Anuta, A. Tudose, C. Celia, D. Paolino, M. Fresta, R. Sandulovici and I. Mircioiu. 2019. Mathematical modeling of release kinetics from supramolecular drug delivery systems. *Pharmaceutics*, 11(3), 140.

Piluharto, B., V. Suendo, I. Maulida and A. Asnawati. 2017. Composite beads of chitosan/bentonite as a matrix for phosphate fertilizer controlled-release. *Journal of Chemical Technology and Metallurgy*, 52(6), 1027-1031.

Pulimi, M. and S. Subramanian. 2016. Nanomaterials for soil fertilisation and contaminant removal. In: *Nanoscience in Food and Agriculture* 1 (pp. 229–246). Springer, Cham.

Rahman, M.H., K.S. Haque and M.Z.H. Khan. 2021. A review on application of controlled released fertilizers influencing the sustainable agricultural production: A cleaner production process. *Environmental Technology & Innovation*, 23, 101697.

Roy, A., S.K. Singh, J. Bajpai and A.K. Bajpai. 2014. Controlled pesticide release from biodegradable polymers. *Central European Journal of Chemistry*, 12(4), pp. 453–469.

Singh, B., D.K. Sharma, R. Kumar and A. Gupta. 2010. Controlled release of thiram from neem-alginate-clay based delivery systems to manage environmental and health hazards. *Applied Clay Science*, 47(3–4), pp. 384–391.

Tan, D., P. Yuan, F. Annabi-Bergaya, F. Dong, D. Liu and H. He. 2015. A comparative study of tubular halloysite and platy kaolinite as carriers for the loading and release of the herbicide amitrole. *Applied Clay Science*, 114, pp. 190–196.

Vega-Vásquez, P., N.S. Mosier and J. Irudayaraj. 2020. Nanoscale drug delivery systems: From medicine to agriculture. *Frontiers in Bioengineering and Biotechnology*, 8, 79.

Yamamoto, C.F., E.I. Pereira, L.H. Mattoso, T. Matsunaka and C. Ribeiro. 2016. Slow release fertilizers based on urea/urea–formaldehyde polymer nanocomposites. *Chemical Engineering Journal*, 287, pp. 390–397.

Nanosensors for Plant Disease Diagnosis

Sensing technology has been widely utilized in many industrial facets to identify changes in the environment. Sensors are fundamentally applied for process control, monitoring, and safety of a given industrial process. The intrusion of nanotechnology has significantly enhanced the performance and efficiency of sensor systems. Nanosensors are chemical sensors that perceive or measure the presence of a given element or substances on a nanoscale with at least one of their sensing dimensions within the 1–100 nm range. This has been a rapidly emerging field where an augmentation of nanotechnology enhances the selectivity, sensitivity, and speed of detection in comparison to conventional sensing approaches. Currently, these are used in monitoring physical, chemical and biological sensations in regions which are difficult to reach. Also, this has been used to detect biochemical and nanoscale particulates in industry and the environment (Abdel-Karim et al. 2020, Noah 2020).

One aspect of sensing technology is to analyze biological components in an environment. Power sensing technology with nanotechnology enables it to analyze biological components even to a pinpoint level accurately. Nanobiosensors are one of the sensing categories which have been reported to analyze biological components. It is noteworthy to mention that; this is not a special or different type of sensor that utilizes nanomaterials to enhance the efficacy of biosensors. This allows us to detect different metabolites, enzymes, DNA, antibodies, viruses and other microbial cells such as bacteria, fungi or even different compounds secreted by them. Globally there is an exceptional improvement in the research related to nanobiosensors proving their immense impact on society. Therefore, this has been applied in medicine for the detection and diagnosis of pathogens, environmental monitoring, and food and agriculture as well.

Different microorganisms including bacteria, fungi and viruses can infect plants to cause diseases. This can significantly influence the overall health and upset the productivity of crops in agriculture. This has been one of the major challenges in commercial agriculture which leads to greater economic losses for farmers. The severity of the plant diseases by different phytopathogens can be influenced by plant type, the growth stage of the plant, environmental conditions, and co-infections by different pathogens as well as by their virulence. Infection by phytopathogens leads to causing biogenic stresses to the crop plants thus affecting their productivity. Once infected a plant is difficult to control. Therefore, continuous monitoring of crop vigor and early detection of such plant diseases is crucial to lessen the disease spread and crop damage. As precautions are better than cure or remedy, the application of sensors technology as a smart approach has been introduced for effective crop management practices. In contrast to conventional sensors, nanosensors or nanobiosensors have several beneficial assets, such as high sensitivity and selectivity, near real-time detection, low cost and portability.

5.1   Conventional Approaches and their Limitations

Conventional strategies to detect phytopathogens include a visual examination, isolation and culturing of them under selective growth media under aseptic conditions, bioassays which imply specific indicators to inoculate and inspect for specific symptoms, visualization using microscopic techniques including light microscopy and electron microscopy, serological techniques such as immunodiffusion tests, enzyme-linked immunosorbent assays (ELISA), radioimmunosorbant assays, dipstick immunoassays, western blot, tissue blot immune assays. In addition, molecular assays including polymerase chain reaction (PCR), reverse transcription PCR (RT-PCR), quantitative real-time PCR (qtRT-PCR), DNA fingerprinting, isothermal amplification assays, microarrays techniques and nucleic acid blot assays are also employed to detect plant pathogens (Mathivanan 2021).

Lateral flow devices are used to diagnose pathogens by detecting antigens of pathogen origin. The implication of monoclonal antibodies (an antibody produced by a single clone of cells or cell lines and consisting of identical antibody molecules) in serological techniques such as ELISA provide high specificity and selectivity to a target of interest. Furthermore, it has been employed in the on-site field-level identification of pathogens from a large number of crops. Nucleic acid-based assays are steadier and specific for detecting single targets from complex mixtures containing the genomes of multiple pathogens. The PCR is a promising technique to detect single targets in very complicated samples as well. It is capable of identifying uncultivable pathogens from biological samples. Also, advanced PCR techniques such as multiplex-PCR approaches have been developed to detect multiple targets from field samples.

The main burden or the limitations of these approaches is their difficulty to reach conclusions due to fieldwork. For instance, cross contaminations from chemicals and reagents during PCR assays can impose negative results or false positives from the non-targets restricting the application of PCRs under field conditions. Almost all these techniques are reliable mostly at symptomatic stages. In addition, conventional approaches are time-consuming, require costly equipment, and also require professional experts. Therefore, reliable and timely detection of phytopathogens plays a crucial role in crop health monitoring to minimize the disease spread and facilitate effective crop management practices.

At present products have been developed and have used sensor technology to diagnose plant pathogens under field conditions. The use of nanobiosensors in the association of different biomarkers enables the extraction of rapid, precise and accurate information related to the early detection and progress of plant diseases under field conditions. Nanobiosensor offers a new dimension to plant disease diagnosis by offering non-destructive, minimally invasive, economical, and easy-to-use systems. Also, the use of nanotechnology has enhanced the minimum level of detection and sensitivity, and specificity of detecting plant pathogens. These tools are used to solidify cultivations and are more sustainable and safer by minimizing the expenses incurred on agrochemicals to maintain crop health.

5.2 Components of Nanosensors Used in Plant Disease Diagnosis

Nanosensor systems are developed using different nanostructured materials such as porous silicon and carbon-based nanomaterials, metal and metal oxide-based nanomaterials, nanoprobes, nanowire nanosensors, nanosystems such as cantilevers and nano-electromechanical systems (NEMS), and also using polymer nanomaterials. Generally, sensing systems have three basic components namely, receptor system, transducer system and detection system.

Typical biosensing systems include a bioreceptor to recognize bioanalytes, a transducer system to convert signals generated from the interaction of analytes with the receptor to a readable form and finally the detector system which amplifies electrical signals from the transducer components to read and study properly. Figure 5.1 below shows schematically a typical nanobiosensor. Receptors are critical and the most important components of a sensor. These are preferable

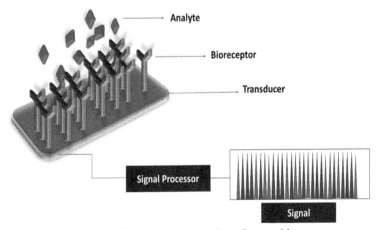

Figure 5.1: Schematic representation of a nanobiosensor.

platforms for the development of chemical or biological sensors to select and identify a target analyte. Receptors in nanobiosensors for the detection of plant pathogens imply antibodies, proteins, enzymes, nucleic acids such as DNA, and their volatile compounds as sensing receptors. It is important to note that the immobilization of bioreceptors using nanotechnological approaches makes their interactions more efficient and feasible. Biological components are complex and they directly associate with the environment and its changes. Nanotechnology has enabled the enhancement of the sensing phenomena of sensing devices. Sensing and identifying changes in dynamic environments and corresponding alterations in its homeostasis at in vivo and ex vivo levels are assumed to be crucial. The development of efficient biological sensors can analyze even minute details of biological interactions.

Another key component of a sensing device is its transduction system which is responsible for transforming responses or signals from bioanalyte interactions to a readable and detectable form. This could be simply the conversion of biochemical energy from a reaction to an electrically readable form through transduction mechanisms. Augmenting nanotechnology into sensing technology has enhanced the potential of sensors to monitor biological phenomena at the physiological levels with far greater precision. The high surface area to volume ratio and the reactivity of nanomaterials allows them to be used in a better and far more diversely functional manner. In addition, the integration of nanomaterials with electrical systems makes nanoelectromechanical systems, which are very active in their electrical transduction mechanisms. For instance, optoelectronic properties of metal-based nanoparticles are very excellent materials for electronic and optical applications. Hence, this can be efficiently used for the detection of nucleic acid sequences (Naresh and Lee 2021).

5.3 Classification of Nanosensors

Based on the sensing mechanism nanosensors can be classified either as mechanical nanosensors or chemical nanosensors. Nanosensors capable of detecting chemicals such as indicator molecules can change the conductivity of the nanomaterial once an analyte interacts with the receptor is detected. Upon binding or adsorption of different analytes onto the receptor reduce the electrical conductivity of nanomaterials in the sensor. Same as chemical nanosensors, mechanical nanosensors also operate under the same principle. However, the mechanisms involved in changing the electrical conductivity differ from chemical nanosensors. In addition, nanosensors can be classified according to the type of receptor molecules incorporated in the sensing system where nanosensors are grouped as affinity-based nanosensors and catalytic-based nanosensors. Furthermore, they can be grouped into several categories based on the structure of the sensor as optical nanosensors or electrochemical nanosensors. Moreover, based on their application nanosensors can be classified as nanobiosensors, chemical nanosensors and electrometers etc. In addition, nanosensors can be classified based on their source of energy, structure and applications (Abdel-Karim et al. 2020, Naresh and Lee 2021).

5.4   Nanoparticles as Sensing Elements

Nanoparticles and nanocomposites alone or in combination with other markers are used in fabricating sensing tools. These are used to detect pathogens directly or to detect indicative compounds involved in disease advancement. For instance, salicylic acid is an important plant hormone known to mediate the host response upon pathogen infection. During an infection, the level of salicylic acid is reported to increase with viral, fungal, bacterial, phytoplasma and insect infections (Lefevere et al. 2020). Moreover, semiconductor metal oxide nanoparticles such as SnO_2 and TiO_2 are used to detect disease-marking volatile organic compounds such as p-ethylguaiacol in strawberry fungus (*Phytophthora cactorum*) which are of pathogen origin. Nanomaterials allow the detection of pathogens easier to perform and able to diagnose infected plant samples in a short period. Table 5.1 here summarizes different types of nanomaterials used to develop sensors and their key benefits.

Fabrication of nanosensors has two vital operations namely, the manufacturing and designing of nanoscale adhesive surfaces via the technology of integrated circuits and the engineering of nanomaterial

Table 5.1: Nanomaterials and their key benefits for sensor technology

Nanomaterial used	Key benefits
Carbon nanotubes	Improved enzyme loading, higher aspect ratios, ability to be functionalized, and better electrical communication
Nanoparticles	Aid in immobilization, enable better loading of bioanalyte, and possess good catalytic properties
Quantum dots	Excellent fluorescence, quantum confinement of charge carriers, and size tunable band energy
Nanowires	Highly versatile, good electrical and sensing properties for bio- and chemical-sensing; charge conduction is better
Nanorods	Good plasmonic materials which can couple sensing phenomenon well and size tunable energy regulation, can be coupled with micro-electromechanical systems, and induce specific field responses

Source: Malik et al. 2013

surfaces through the process of micromachining. This technique, thus developed for biosensing, uses the variations of four basic processes, namely, photolithography, thin film etching/growth, surface etching strategies, and chemical bonding parameter (Noah 2020).

Optical properties of nanoparticles such as Surface Plasmon Resonance can be used to optimize the sharp precise optical responses of sensing materials with the incident light. Surface Plasmon Resonance is the collective excitation of conduction band electrons surface ionic species and charged particles that are in resonance with the oscillating electric field of the incident light, which will produce energetic plasmonic electrons through non-radiative excitation. This property gives them photonic character and excellent ability to be used as fluorophores for nanosensors (Malik et al. 2013, Noah 2020). Table 5.2 depicts the different types of nanosensors applied for the detection of plant pathogens.

Table 5.2: Types of nanosensors applied for the detection of plant pathogens

Sensor	Nanomaterial	Target pathogen	Target	Test host(s)	Detection limit
Electrochemical	Colloidal Au nanoparticles	*Pseudomonas syringae*	DNA	*Arabidopsis thaliana*	
Electrochemical	SnO_2 and TiO_2 nanoparticles	*Phytophthora cactorum*	p-ethylguaiacol	Strawberry	35–62 nM
Fluorometric	Tioglicolic acid capped Cadmium-telluride quantum dots (QDs) and CNPs	Citrus tristeza virus	Antibody	Citrus	220 ng mL^{-1}
Fluorescent Resonance Energy Transfer (FRET)	Tioglicolicacid-modified Cadmium-telluride QDs (CdTe-QDs)	*Polymyxa betae*	Antibody (Anti glutathione-S-transferase protein)	Beta vulgaris and Hordeum vulgare	0.5 lg mL^{-1}
FRET	CdSe QDs	*Ganoderma boninense*	DNA	Oil palm	1.12×10^{-12} M
FRET	Tioglicolic acid-modified cadmium-telluride quantum dots (CdTe-QDs)	Candidatus Phytoplasma aurantifolia (Ca. *P. aurantifolia*)	Antibody (IMP)	Lime	5 ca. *P. aurantifolia*/ μL^{-1}

Piezoelectric	Gold NP-coated QCM crystals	Maize chlorotic mottle virus	Antibody (anti-MCMV)	Maize	250 ng mL^{-1}
Piezoelectric	QCM crystals	Cymbidium mosaic potexvirus (CymMV) and odontoglossum ringspot tobamovirus (ORSV)	DNA	Orchid	1 ng
Surface Enhanced Raman Scattering (SERS)	Nanotages composed of Au NPs core	Botrytis cinerea, Pseudomonas syringae, and F. oxysporum f. sp. conglutinans, F. oxysporum f. sp. lycopersici	DNA	Arabidopsis thaliana and tomato	2.5 ng
SERS	Ag NPs	Phytophthora ramorum	DNA	Rhododendron	

Source: Kashyap et al. 2019

5.5 Potential Applications of Nanosensors in Disease Diagnosis by Phytopathogens

It is reported that piezoelectric sensors can be a better choice for the rapid, precise and reliable diagnosis of phytopathogens under field conditions. Nanobiosensor-based assays require only about 20 microliters of the biosamples and are useful in critical situations where samples are limited and fewer in quantity. Carbon nanomaterials are the best materials for the development of fluorometry-driven nanosensors. FRETs-based nanosensors in association with quantum dots pose unique properties that can facilitate in designing and fabricating of novel and better-quality sensors under field applications. FRET nanosensors comprised of fluorophore pairs having overlapping emission spectra will allow the detection of conformational changes that occur during the energy transfer between fluorophores. Analyzing the emission peaks of acceptor and donor molecules facilitates the ratiometric detection of different target molecules. There are two main categories of FRET nanosensors. Firstly, as genetic constructs within the plants such as fluorescence proteins. Secondly, externally synthesized products such as nanoparticles. Although this is a better nanosensor approach to detect plant pathogens, photobleaching and the lower stability of the fluorophores used creates practical issues in utilization.

Surface-enhanced Raman scattering (SERS) based sensors largely depend on DNA-DNA hybrid formation and could be performed in portable field Raman devices. This will allow rapid and on-the-spot detection of pathogens under field conditions. SERS- labelled nanotags are used to detect multiple pathogens causing infections in crops and these are more sensitive than conventional PCR. Also, this can detect as few as two copies of pathogenic DNA. In addition, metal oxide nanoparticles provide reasonable and economical alternatives to costly electrode materials like gold and platinum in designing amperometric sensors.

Nanobiosensors can be utilized in the detection of soil pathogens such as fungi, bacteria and viruses. These sensor systems diagnose soil diseases by quantifying the oxygen utilization by soil microorganisms during their respiration. Detection and diagnosis using these sensors involve the following steps. Firstly, two sensors saturated with good and bad microbes will be selected separately and drenched in soil suspension. Next, the sensors will evaluate the oxygen consumption by microbes and data will be recorded for further analysis. Finally, data obtained from the two groups of microbes will be compared to understand the best group of microbes suited for the soil (Omanović-Mikličanina and Maksimović 2016).

Furthermore, nanobiosensors can be used in the detection of seed storage conditions. Under storage conditions, the aging seeds tend to release different volatile aldehydes. These volatile organic compounds can cause detrimental effects on other seed types. Therefore, it is imperative to detect volatile aldehydes in the storage area and to separate damaged and unviable seeds from fresh ones' electronic noses made using nanomaterials can be used. These nanosensors can identify and group different types of vapours, gases and odours routinely (Ghaffar et al. 2020, Mathivanan 2021).

5.6 Conclusion and Future Perspectives

Nanotechnology as an interdisciplinary approach made several modifications to sensor technology enhancing the efficiency of its applications. Therefore, it has been used to improve the quality of the environment, economy, agriculture, medical field and even in the food industry. Integration of nanomaterials on biosensors enhances their sensitivity immensely and they are easy to utilize. Furthermore, they require lesser quantities of samples than conventional approaches. Also, nanobiosensors can be portable and engineered to have faster detection rates. Unique physicochemical properties of nanomaterials enable measurement of a wide range of variables in the environment. Hence, it allows the detection and identifies different phytopathogens damaging crops by diagnosing the target organisms directly or indirectly by detecting different target molecules and compounds. Nanosensors involve binding or reaction with target species or strains and eventually transform these interactions into detective signals, thereby facilitating quick, precise and early diagnosis of disease-causing agents to ensure the quality of food production. In addition, real-time detection and analysis of environmental conditions are important aspects of current agriculture specially in smart and sustainable agricultural practices. Although, this has been promised in many applications including the potential plant disease diagnosis still nanobiosensor technology under the initial phase and a dearth of nanoscale quantification. Hence, still there is a need for advanced optimizing sensing elements for a sustainable and efficient agricultural system specifically for the sustainable use of resources and detection and diagnosis of crop and field stress and better post-harvest care. Also, basic research is still required to improve the present-day sensor technology, sensing strategies and the refining of analytical instrumentations and protocols for crop disease diagnosis at an early stage.

References

Abdel-Karim, R., Y. Reda and A. Abdel-Fattah. 2020. Nanostructured materials-based nanosensors. *Journal of the Electrochemical Society*, 167(3), 037554.

Ghaffar, N., M.A. Farrukh and S. Naz. 2020. Applications of nanobiosensors in agriculture. In: *Nanoagronomy* (pp. 179–196). Springer, Cham.

Kashyap, P.L., S. Kumar, P. Jasrotia, D.P. Singh and G.P. Singh. 2019. Nanosensors for plant disease diagnosis: Current understanding and future perspectives. In: *Nanoscience for Sustainable Agriculture* (pp. 189–205). Springer, Cham.

Lefevere, H., L. Bauters and G. Gheysen. 2020. Salicylic acid biosynthesis in plants. *Frontiers in Plant Science*, 11, p. 338.

Omanović-Mikličanina, E. and M. Maksimović. 2016. Nanosensors applications in agriculture and food industry. *Bull. Chem. Technol. Bosnia Herzegovina*, 47, pp. 59–70.

Malik, P., V. Katyal, V. Malik, A. Asatkar, G. Inwati and T.K. Mukherjee. 2013. Nanobiosensors: Concepts and variations. *International Scholarly Research Notices*, 2013.

Mathivanan, S. 2021. Perspectives of nano-materials and nanobiosensors in food safety and agriculture. *Novel Nanomaterials*, 197.

Noah, N.M. 2020. Design and synthesis of nanostructured materials for sensor applications. *Journal of Nanomaterials*, 2020.

Naresh, V. and N. Lee. 2021. A review on biosensors and recent development of nanostructured materials-enabled biosensors. *Sensors*, 21(4), p. 1109.

Nanomaterials in Seed Germination and Plant Growth

Seed germination can be defined as the process by which a plant grows from a seed into a mature plant and it initiates with it the hydration of the seed and then terminates following the emergence of the embryonic axis. Seed germination at the right time promotes the development of a deep root system that enables the plants for their proper establishment, whereas slow germination can make young seedlings prone to various biotic and abiotic stresses. When this phenomenon occurs rapidly and homogeneously, it facilitates the successful establishment of plants with a deep root system. In many instances, seed germination reduces due to various environmental stress factors.

6.1   Physiology of Seed Germination

A prerequisite for seed germination is the absorption of water into the seeds. Due to the absorption of water, a complete alteration of physiology takes place inside the cells. Imbibition is a type of diffusion during which the water (imbibate) is absorbed by colloid-type solid particles called imbibants resulting in increasing the volume. The capacity of different imbibants differs due to their physical and chemical characteristics. For instance, if compared to proteins and starch, starch has less capacity as an imbibant to absorb water. The water potential gradient between the imbibate and imbibant, pH of the medium and texture of the imbibant and affinity (attractive forces) that work between both imbibant and imbibate and temperature are the influential factors of imbibition. Three phases involve the entry of water into the seeds through imbibition.

At first rapid uptake of water initiates the basal metabolism in seeds. During Phase I this starts mechanisms related to transcription, protein synthesis, and mitochondrial activity. Secondly the phase of enhanced water uptake and finally the uptake of water along with the initiation of growth. The changes that occur during the second phase cause breaking the dormancy even under adverse conditions. During prolonged water stress conditions, the mitochondria get disassembled which results in a reduced respiration rate while cells get disintegrated. Importantly in Phase II, several events that prepare the seeds to be germinated take place. Accordingly, repairing the damaged mitochondria with an increased rate of respiration, repairing DNA through base and nucleotide excision pathways and up-regulation of germination-associated genes due to which the synthesis of related proteins takes place. Imbibition helps expand cells that are driven by pressure potential and increases germination. Moreover, imbibition itself damages cells due to the rapid absorption of water. In Phase III, faster absorption of water takes place similar to Phase I. Finally, the emergence of the radicle ends the growth of the embryo.

The process of seed germination finally encompasses the activity of plant hormones and reactive oxygen species (ROS). ROS generation is directly linked with the apoplastic pathway of water transportation through cell walls. Cell wall loosening facilitates the uptake of water and increases of cellular length. The activity of ROS, when present at optimal concentrations, regulates gene expression and signaling mechanisms of plant growth regulators (Carrera-Castaño et al. 2020, Do Espirito Santo Pereira et al. 2021).

6.2 Nanomaterials for Enhancing Seed Germination and Seedling Emergence

6.2.1 Seed Nanopriming

Seed priming is considered one of the most widely practiced methods that enhance seed germination. Priming involves mixing of seeds with a specific solution at a known concentration for a specific period that ultimately alters the physiology of seeds. Water is the simplest and most commonly used reagent for priming since ancient times and substances that are soluble or suspendable in water too are being used (growth regulators, nutrients, biopolymers, microorganisms and salts). Priming can also be coupled with alterations in temperature as a measure to increase the efficiency of the process.

Priming results in speedy and timely germination even under adverse conditions. The importance of this technology exists because it can enhance the germination of weak, damaged and old seed lots. Priming also helps in increasing seed quality and seedling establishment with enhanced plant growth with a uniform population that increases the yield. Together with these major benefits, this also increases plant tolerance to stresses caused by both biotic and abiotic factors and increases water use efficiency, nutrient uptake and nutritional quality of crops (Sen and Puthur 2020).

Nanopriming has shown to be effective because of the unique physical and chemical properties of materials at the nanoscale. It involves the priming of seeds with nanomaterials that are in a suspension or formulation. Here the nanomaterials do not essentially require to be up taken, while even in the case of uptake, minute amounts are being up taken.

Application of nanotechnologies in seed priming which is called nanopriming is in two ways namely, active nanoparticles and nanocarriers incorporated with suitable nanoparticles having sustained release properties. Here, the active nanoparticles that are synthesized either by chemical or biogenic synthesis processes exert a positive effect on plant growth and development, whereas nanocarrier-mediated systems release nanomaterials in a sustained manner. In most instances, the active nanoparticles are synthesized by essential elements that play vital roles in plant systems. The nanomaterials in nanocarrier systems can be active or inactive or may be loaded with an active ingredient. In these systems, certain biopolymeric nanoparticles derived from polysaccharides, lipids, and proteins have widely been researched, e.g., natural biopolymers like alginate, lignin, cellulose, synthetic biopolymers like chitosan, and lipid nanoparticles. Low concentrations of these polymeric nanoparticles exert positive effects on seed germination. Moreover, metal nanoparticles can be incorporated into polymeric nanocarrier systems. Moreover, nanodelivery systems can be used as multipurpose systems. For instance, delivery systems targeting the controlled delivery of agrochemicals can also be incorporated with nanomaterials that can be employed for seed priming purposes.

Research has proven the ability of nanoprimed seeds in increasing the dry matter content, vigor and also morphological traits. Moreover, it triggers certain physiological processes that take place during the process of seed germination. Cell growth, repair and proliferation and photosynthesis are such processes that enhance through nanopriming. In addition to the above, nanopriming enhances the potential to withstand pest and disease infestations. Varied nanoparticles that are either metallic or non-metallic in origin have been researched in terms of nanopriming that showed promising positive effects. Accordingly, SiO_2, FeO, Fe_2O_3, TiO_2, Au, Ag, Zn and CuO have shown positive effects on seed germination. In contrast, the same nanoparticles have shown negative effects on seed germination and seedling growth at different concentrations. Consider whether the effects are positive or negative depending on the nanomaterial species, its concentration, plant species and seed size purpose (Do Espirito Santo Pereira et al. 2021, Fiol et al. 2021).

6.3 Revitalizing Aged Seeds

Aging is a natural and time-dependent phenomenon that cannot be inverted. The result will have deteriorated seeds. Hence, measures to slow down the rate of aging through the regulation of storage environmental factors will be the only remedy to overcome this challenge.

Aged seeds contribute to the retarded germination of seeds. Seed aging is associated with the activity of reactive oxygen species (ROS) because their cells can initiate processes that increase ROS generation and neutralize antioxidants. This phenomenon can also reduce the germinability of seeds. Conventionally, when storing seeds, ROS blockers are used to reduce aging. Also, storing seeds in hydrated storage facilities is another method used to reduce aging. Nanotechnologies have been proven to be a novel approach to revitalizing and reviving aged seeds. Accordingly, some inorganic nanoparticles have been shown to be working on neutralizing the excess ROS and initiating signals generated by the growth regulators to support early emergence.

Certain legume species such as chickpea, long beans, soybean, and ground nut have been tested by incorporating both metallic and non-metallic nanomaterials in terms of their potential to reduce ageing. Not only legumes but also certain other food crops such as corn and rice have also been tested and proven the effectiveness in terms of the same.

In general, most of the research studies conducted have employed nanomaterial concentrations ranging between 400–1500 mg/kg that has shown positive effects, however when the concentration increases higher than 1000 mg/kg reduction in efficacy could be observed. Biogenic nanomaterials are effective to overcome the effects of aging of seeds. Such nanomaterials are inherently possessed functional groups or else can be coated with natural reducing agents that work on reducing the ROS levels in seeds.

6.4 Genetic Regulation of Increasing the Germinability due to Nanopriming

Several genes that are associated with enhancing germination or inducing dormancy can be found in plant genomes. Upregulation of genes including the Mn-superoxide dismutase gene (Mn-SOD gene) involved in increasing SOD that protects cells caused by ROS damage. Also, the Phenylalanine ammonia-lyase gene (*PAL* gene) is involved in working on the innate immune system and pathogenicity-related genes (*PR* genes) are involved in the salicylic acid synthesis. In addition to that, the expression of microRNA has been demonstrated upon exposure to nanomaterials (Do Espirito Santo Pereira et al. 2021).

6.5 Nanomaterials in Plant Growth

Plant growth can be discussed at three different levels viz. meristem level which involves proliferation of cells that contribute to organ development, at organ or system level that involves tissue expansion and finally in long term increase of biomass. Overall, it includes an irreversible increase in plant biomass with time and can be divided into three phases namely, lag phase, exponential phase and stationary phase.

Lag Phase

In this phase, the rate of growth is very slow. During this phase, the cell size increases and cells become metabolically active.

Exponential Phase

In this phase, the number of cells increases by following a logarithmic pattern. The growth rate becomes highest and with a higher rate compared to the lag phase. The physiological activity of the plants exists at a maximum level during the exponential phase.

Stationary Phase

Stationary phase is also referred to as an adult phase or final steady state of growth. In the stationary phase, the growth rate reduces due to the limitations of nutrients and accumulation of toxic by-products, while the physiological activity of cells also retards. This phase denotes the maturity of the growth system.

Plants require essential elements to continue their growth and development. In this context, about seventeen elements that include both metallic and non-metallic elements have been documented. Some of these elements are good, while some are poor in terms of the mobility inside the plants and these elements regulate functions such as photosynthesis, respiration, phloem translocation, cell structure, division and elongation, enzyme activity, and metabolism of various compounds. Nanomaterials that are derived from some of the essential elements have also been synthesized and their effects have been documented after adding to the plants. ZnO, CuO, MgO, and iron oxide nanoparticle species including FeS_2, FeO, Fe_2O_3, Mn and NiO have been tested for the effects. In addition to those certain other nanomaterials are derived from non-essential elements such as Ag, Au, TiO_2, and carbon nanotubes (SWCNT and MWCNT). These nanomaterials have been applied to plants as foliar sprays, mixing with growth substrate or drenching or incorporated into the hydroponic solution. On an industrial scale, the incorporation of nanomaterials for the growth and development of plants is a better way to overcome the challenge of nutrient use efficiencies of conventional fertilizers. Together with that, drawbacks associated with conventional nutrient formulations including retention in soil and flowing through the surface and groundwater the world is thinking more than twice about continuing agriculture with these conventional fertilizer formulations.

6.6 Metallic Nanomaterials

One of the widely used nanomaterial species is metal nanoparticles. ZnO, FeS, FeO, Fe_2O_3, CuO, Ag, and TiO_2 are some of the widely researched metal nanoparticle species used to conduct plant-based studies. ZnO which is one of the widely used nanoparticle species has shown enhancement of growth, biomass, net photosynthesis, nutrient content, grain yield, metabolites, enzyme activity and structural properties in plants, while FeS has increased crop yield and Fe_2O_3 has shown plant growth and biomineralization. CuO could also increase biomass production with increased photosynthetic efficiency and finally, TiO_2 could induce biomass production and chlorophyll and N contents within the plants (Ali et al. 2021).

6.7 Non-metallic Nanomaterials

Carbon nanomaterials: carbon nanotubes (CNT), carbon dots, fullerenes (C60) and fullerols $[C_{60}(OH)_n]$ are discussed hereunder in terms of their effects on seed germination and plant growth. *Brassica juncea* could increase root and shoot length more than 1.5 times when treated with CNT compared to the plants that were not treated. Moreover, Broccoli plants also showed positive effects on growth when exposed to CNT. In tomatoes, CNT treatment could enhance flower production by twice the non-treated plants. An increase in the lengths of roots and stems, biomass accumulation, root vigor, single plant fresh weight and increasing carbohydrate content and photosynthesis have been observed in mung beans when treated with carbon dots. Moreover, records exist on enhanced growth of *Brassica parachinensis* L. and *Lactuca sativa* while in coriander and garlic plants, including leaves, roots, shoots, flowers and fruits, the growth rate was observed to be increased. Furthermore, in wheat and rice plant when exposed to carbon dots, the root and shoot length was increased. Fullerenes can enhance the uptake of certain nutrients into the plants, for instance, nitrogen has been well absorbed e.g., teak (*Tectona grandis*) seedlings while increasing the nutrient content and growth of teak plants. Also, fullerenes showed stimulatory effects on root growth in barley. Moreover, SiO_2 could improve the growth and grain weight of wheat when under drought stress (Shang et al. 2019, Li et al. 2020, Patel et al. 2020).

6.8 Co-application of Synthetic Nanomaterials

Co-application refers to the application of one or more nanomaterial species or nanomaterials with another substance. Co-application of synthetic nanomaterials that are favorable for the growth of plants has been demonstrated in many research studies. TiO_2 incorporated with plant growth promoting bacteria (PGPB) have enhanced the growth of white clover plants. Furthermore, other nanoparticle species such as CuO, SiO_2, SiO_2 and Au, Fe, Zn and Mn when incorporated into plants with PGPB have shown enhanced growth by increasing the root, shoot lengths and soil nutrient content.

6.9 Conclusion and Future Perspectives

Nanomaterials possess positive effects on germination, growth and development of plants under optimum concentrations. Setting benchmarks for these optimum concentrations is however a challenge. This is because the optimum concentrations vary with the crop species and growth stages. This nature in turn limits the nanomaterials being applied to a wide range of crop species and at different growth stages. The magnitude of the problem increases when it comes to commercial formulations. Hence, having a better understanding of the effects of nanomaterials per se on growth and development plus when the nanomaterials are at the formulation is also important to consider. Meanwhile having better research information based on the co-application of nanomaterials with other substances is also a potential research area to be investigated.

References

Ali, S., A. Mehmood and N. Khan. 2021. Uptake, translocation, and consequences of nanomaterials on plant growth and stress adaptation. *Journal of Nanomaterials*, 2021, pp. 1–17.

Carrera-Castaño, G., J. Calleja-Cabrera, M. Pernas, L. Gómez and L. Oñate-Sánchez. 2020. An updated overview on the regulation of seed germination. *Plants*, 9(6), p. 703.

Do Espirito Santo Pereira, A., H. Caixeta Oliveira, L. Fernandes Fraceto and C. Santaella. 2021. Nanotechnology potential in seed priming for sustainable agriculture. *Nanomaterials*, 11(2), p. 267.

Fiol, D.F., M.C. Terrile, J. Frik, F.A. Mesas, V.A. Álvarez and C.A. Casalongué. 2021. Nanotechnology in plants: Recent advances and challenges. *Journal of Chemical Technology & Biotechnology*, 96(8), pp. 2095–2108.

Li, Y., X. Xu, Y. Wu, J. Zhuang, X. Zhang, H. Zhang, B. Lei, C. Hu and Y. Liu. 2020. A review on the effects of carbon dots in plant systems. *Materials Chemistry Frontiers*, 4(2), pp. 437–448.

Patel, D.K., H.B. Kim, S.D. Dutta, K. Ganguly and K.T. Lim. 2020. Carbon nanomaterials and their agricultural and biotechnological applications. *Materials*, 13(7), p. 1679.

Sen, A. and J.T. Puthur. 2020. Seed priming-induced physiochemical and molecular events in plants coupled to abiotic stress tolerance: An overview. *Priming-Mediated Stress and Cross-Stress Tolerance in Crop Plants*, pp. 303–316.

Shang, Y., M. Hasan, G.J. Ahammed, M. Li, H. Yin and J. Zhou. 2019. Applications of nanotechnology in plant growth and crop protection: A review. *Molecules*, 24(14), p. 2558.

Toxicity of Nanomaterials on Plants and their Habitats

Nanomaterials can have a positive, negative, or neutral impact on plant growth and development. However, the mechanisms behind these effects are not fully understood, and thus need to be explored further. In general, under lower concentrations with a size less than 50 nm, applied nanomaterials can trigger various structural and physiological parameters of plants such as seed germination, root length, shoot length, leaf area, photosynthesis, and dry biomass in plants like wheat, rice, maize, and onion. The ability of different nanomaterials to be absorbed into the plant systems at varied efficacies and their effects on the nutrient absorbance mechanisms is the reason behind this phenomenon.

7.1 Effects of Nanomaterials on Nutrient Uptake by Plants

The effects of nanomaterials on nutrient uptake can occur in two ways: either by triggering nutrient uptake or by directly or indirectly retarding nutrient uptake. Direct uptake involves providing the nutrient element itself through the nanomaterial added, e.g., Zn^{2+} from ZnO nanoparticles, while indirect uptake involves the enhancement of the bioavailability of certain other elements (Adele et al. 2018). Bioavailability is simply, the capacity of the soil-plant system to supply and facilitate absorption of nutrients into plants during a specific period if the plants are grown in a soil medium.

Accordingly, soil nutrient bioavailability depends upon three major variables namely:

1. The ability of nutrients to be released from the solid phase to the solution phase
2. The mobility of nutrients from soil solution to the root mycorrhizae
3. Absorption of nutrients by the plant root–mycorrhizal system (Judy et al. 2016, Noori et al. 2017).

For instance, CuO can regulate the uptake of macronutrients such as P, and S and also can trigger the uptake of micronutrients such as Cu itself and Na. Moreover, ZnO can trigger the uptake of macronutrients such as Mg and Ca and micronutrients such as B and Mo. In addition, TiO_2 could enhance the uptake of certain micronutrients such as Fe, Cu, and Mn depending on the surface characteristics of the nanoparticles. Nanomaterials can enhance the uptake of both macronutrient and micronutrient elements. Therefore, the addition of nanomaterials to plants can lead to the absorption of certain nutrients from the soil. Table 7.1 provides some examples of such negative nutrient interactions as a result of adding nanomaterials to the soil.

The soil-plant system's capacity to supply/absorb nutrients is termed soil nutrient bioavailability and are the ability of the soil-plant system to supply essential plant nutrients to a target plant, or plant association, during a specific period as a result of the processes controlling its availability.

7.2 Toxic Effects on Plant Functionality

Toxicity on plants can be simply defined as the toxic effect on plant growth and development caused by any substance called phytotoxins or growth conditions. Accordingly, phytotoxins can affect various processes associated with plants, including seed germination, growth, and development of specific components such as elongation of roots and shoots, change of chemical compositions of cells, etc. Size, shape, surface area, surface charge, chemical composition, agglomeration or aggregation capability, and dissolution ability of nanomaterials are the major determinants that govern the toxicity of

Table 7.1: Retardation of plant nutrient uptake by plants

Plant species	Entry point	Nanoparticles	Nutrient of which the uptake was retarded	
			Macronutrients	**Micronutrients**
Barley	Leaves	Citrate-Au	-	Mn
Basil	Roots	TiO_2-Al_2O_3- dimethicone	P, Ca	Cu
Bean	Shoots	CuO	Ca	Fe, Zn
Bean	Leaves	ZnO	-	Mo
Bell pepper	Leaves and fruits	CuO	-	Zn
Cucumber	Roots	CeO_2	-	Mo
Cucumber	Roots	ZnO	-	Mo
Lettuce	Shoots	CuO	P, Mg, Ca	Mn, Fe
Tomato	Leaves and stems	SnO_2	K, Mg, S, Ca	-
Transgenic cotton	Shoots and roots	CeO_2	P, K, Mg, Ca	Fe, B, Zn, Cu, Mn
Transgenic cotton	Shoots and roots	SiO_2	Mg	Fe, Zn, Cu

nanomaterials. Let us look at how these characteristics affect toxicity in brief. Many studies have shown that when the size dimensions of certain materials are reduced, it can lead to toxic effects compared to their bulk materials.

7.2.1 DNA Damage

DNA damage can occur due to several structural change mechanisms at the molecular level, viz. hydrolytic deamination, alkylating agents, free radicles, and reactive oxygen species. This occurs by interacting DNA with various substances that bind to the DNA noncovalently owing to affecting vital processes such as transcription and translation (García-Sánchez et al. 2021). Entry of nanomaterials to the nucleus occurs via nuclear pores and then directs them towards the nucleus, letting them directly interact with DNA. Reactive oxidation agents can target the most reactive sites of the bases of DNA molecules.

7.3 Effects on Plant Physiological Functions

7.3.1 Effects on Plant Metabolism and Related Mechanisms

Metabolism is the whole range of chemical and physical processes within the cells of living organisms that could happen spontaneously (releasing energy) or non-spontaneously (requiring energy). Metabolism produces metabolites based on their function and biosynthetic pathways. Plant metabolism can be categorized as primary metabolism and secondary metabolism.

Primary metabolites are organic substances that directly deal with biological processes such as growth, development, and reproduction and are produced by primary metabolic pathways. This belongs to diverse compounds, including carbohydrates, amino acids, nucleic acids, lipids, organic acids, and steroids. Plants' primary metabolism is critical for their survival, and secondary metabolism is imperative for the growth and development of plants, including the interaction of the plant with the environment. Primary metabolites are mainly derived from derived glycolysis, the Krebs cycle, or from shikimate pathways for the synthesis of precursor molecules for secondary metabolites. Primary metabolism reactions are highly conserved in plants. However, secondary metabolism pathways are highly diverse and vary from cell to cell at different developmental stages. These metabolic pathways involve the synthesis of carbohydrates, proteins, lipids, and nucleic acids. Nanomaterials show different impacts on plant metabolism, and these can be positive or negative (Hitami et al. 2016, Pott et al. 2019).

Secondary metabolites are synthesized from primary metabolites that are produced by specialized reactions catalyzed by different enzymes. Generally, plant secondary metabolites can be divided into three large chemical classes, namely terpenes that are composed almost entirely of carbon and hydrogen, phenolics that are generally made from simple sugars and have benzene rings, oxygen and hydrogen, and nitrogen-containing compounds such as alkaloids. Typically, secondary metabolites are not directly involved in the normal growth, development, or reproduction of plants. However, they govern vital ecological functions and play imperative roles in protecting plants against various biotic and abiotic stresses.

Nitrogen is vital for the regulation of cellular activities in cells. It is a major component in proteins, DNA, RNA, enzymes, vitamins, hormones, and secondary metabolites such as alkaloids and amides (Hawkesford et al. 2012). Enzymes such as glutamate dehydrogenase and glutamine synthetase are mainly involved in assimilating NH_4^+ into amino acids, which subsequently accumulate as proteins in plant tissues for regulating the basic functions of plants. The potential of nanomaterials to regulate nitrogen metabolism is mainly dependent on the interaction with the following enzymes. For instance, TiO_2 nanoparticles significantly improve nitrate reductase, glutamate dehydrogenase, glutamine synthase, and glutamic-pyruvic transaminase enzyme activities (Abdel Latef et al. 2018, Ahmad et al. 2018). This can promote the nitrate uptake and conversion of inorganic nitrogen into nitrogen-containing organic compounds, which will subsequently enhance the plant biomass. The application of carbon-based nanomaterials such as graphene can up-regulate the biosynthesis of carbohydrates, aromatic amino acids, and fatty acids, leading to improved nitrogen sequestration, secondary metabolism, and oxidation resistance. Also, these nanomaterials are capable of enhancing the asparagine and glutamine content, which play a central role in nitrogen transport and storage in plants.

Furthermore, silver-like nanoparticles can induce cysteine biosynthesis and their related proteins, enzymes, and also reactive oxygen species detoxification pathways. Silver nanoparticles can alter proteins in the endoplasmic reticulum and vacuole. These can be considered potential targets of plant metabolism. Furthermore, the same nanomaterial has been reported to induce oxidative stress and changes in specific cellular functions. This observation indicates that the changes in the levels of enzymes involved in energy metabolism may help cells produce more reducing power to expedite the response to the stress induced by silver nanoparticles. However, the allocation of further energy to promote plant defense systems in root tissues may also contribute to the decreased growth of seedlings exposed to nanoparticles.

7.3.2 Excessive Production of Reactive Oxygen Species and Oxidative Enzyme Activity

Oxidative stress is a phenomenon caused due to imbalance between the production and accumulation of reactive oxygen species (ROS). They act as key signaling transduction molecules as a response to both biotic and abiotic environmental stresses. ROS are produced inside organisms during biochemical processes such as redox chain reactions or as metabolic by-products. ENPs, cause phytotoxicity takes due to excessive production of ROS, which results in protein oxidation, cellular DNA damage, electrolyte leakage, lipid peroxidation, and apoptotic cell death.

To overcome oxidative stress, specific antioxidant defense enzymes quench certain types of ROS through a charge transfer mechanism. As a front line of defense, antioxidative enzymes regulate the levels of ROS by rapid neutralization of free radicles. For instance, catalyzing the dismutation of superoxide radicles into hydrogen peroxide and eventually break down them into harmless molecules. However, during this process, produced hydrogen peroxide initiates metal ion-catalyzed Fenton reactions and can generate highly toxic reactive oxygen intermediate HO (Hydroxyl radicals) which cannot be detoxified with any known antioxidant enzymes.

Indeed, nanomaterials induce positive impacts on certain plants, while the same nanoparticle can cause unfavorable effects on other plants. Such effects are dependent on the intrinsic properties of nanoparticles, the environment, and the plant species. For instance, silver nanoparticles have been shown to down-regulate the expression of proteins like HCF136 protein, required for efficient biogenesis of photosystem II and cytochrome b5, which are involved in the electron transfer chain in photosynthesis. Studies have shown that the exposure of plant roots to silver nanoparticles upregulates the expression of 14-3-3 family proteins. These proteins are capable of binding to different cellular signaling molecules, including protein kinases, phosphatases, and transmembrane receptors. The enhanced level of 14-3-3 proteins acts as an indicator of improved plant defense concerning silver nanoparticle exposure. Another example is that graphene oxide nanoparticles can inhibit the metabolism of amino acids and carbohydrates, reducing the energy supply for the synthesis. This effect can cause mitochondrial respiratory dysfunction triggered by graphene oxide. In addition, graphene oxide nanoparticles increase the ratio of unsaturated to saturated fatty acids, leading to enhanced membrane liquidity and reducing the strength of plant cell membranes. These events show that nanomaterials can severely suppress metabolism in plants.

Studies have shown that all types of nanomaterials can induce oxidative stress and produce excess reactive oxygen species, which can potentially affect protein structure and their levels of expression. Besides, it has been suggested that the impact of nanoparticles on enzymatic proteins depends on the composition, concentration, size, and physicochemical properties of nanomaterials as well as on the plant species. Metal ions released by nanoparticles following an intrusion into the cells may alter protein functionality. The mechanical impact of these nanoparticles depends primarily on their size, rather than their chemical properties. Also, the surface characteristics of nanomaterials have attracted a great deal of attention in the field. Nanomaterials, due to their specific surface characteristics, tend to form layers of OH^- groups at the surface, giving them negativity. This makes the positively-charged side groups of proteins absorb nanomaterials, thus interrupting their normal functions. Nanoparticles can interact with the functional groups of biological molecules using various types of interactions. For instance, functionalized nanomaterials containing various functional groups on their surfaces can react with the functional groups of protein via electrostatic interactions while the bare nanoparticles can make Van der Waals interactions with the surface of the biomolecules.

Sometimes nanomaterials alter the biochemistry of plasma membranes and prevent the expression of proteins associated with ATP production. Therefore, these effects show that nanoparticles can have positive or negative effects, and they can up-regulate or down-regulate the expression pathways for plant metabolism as well.

7.3.3 Effects on Photosynthesis

Plants are considered photoautotrophs, meaning that they can synthesize their food by harvesting light energy. Photosynthesis is an anabolic pathway that is the driving force behind synthesizing their foods. Nanomaterials have both positive and negative effects on photosynthesis. Nanomaterials show a significant impact on photosynthesis efficiency, photochemical fluorescence, and quantum yield. Furthermore, it can affect photosynthetic pigments depending on the species and exposure concentrations (Kaňa 2016).

It has been reported that phytotoxic effects include decreased chlorophyll and carotenoid content, photosystem II efficiency, photophosphorylation, and effects on enzymes that regulate photosynthesis by different forms of nanomaterials. It has been shown that reactive oxygen species generation causes oxidative stress and membrane lipid peroxidation, which leads to these effects. Chlorophyll is considered the main photosynthetic pigment in plants. These pigments are highly sensitive to photodegradation compared to other photosynthetic pigments. Hence, this can be used as a toxicity indicator in plants. For instance, silver nanoparticles are presumed to have a toxic impact on plants by influencing photosynthetic pigments and photosynthetic performance. A reduction in chlorophyll content severely affects photosynthesis. This leads to the release of excess amounts of electrons with molecular oxygen, generating reactive oxygen species. Hence, this can damage the composition of the chloroplast, causing a loss of photosynthesis efficiency. Furthermore, carbon-based nanomaterials such as graphene oxide tend to decrease chlorophyll biosynthesis by damaging the chloroplast structure (Chandra and Kang 2016).

Under high light intensities and nutrients, TiO_2 and CeO_2 nanoparticles have been shown to decrease the photosynthetic rate and CO_2 assimilation efficiency of plants. This is possibly due to the disruption of energy transfer from photosystem II to the Calvin cycle. Furthermore, $Cu(OH)_2$ nanomaterials inhibit photosynthesis in plants grown under severe stress conditions such as high light intensity and nutrient limitation by inhibiting photo-oxidation of the photosystem II reaction center. The effects that are caused by metallic nanomaterials are consequences of the interference of metal ions with photosynthetic enzymes and chloroplast membranes. Furthermore, the accumulation of heavy metals in higher plants affects photosynthesis indirectly by reducing the functioning of the stomata. This also has a significant impact on the transpiration rates of plants as well. Therefore, controlled applications of nanomaterials should be optimized for the direct application of nanomaterials. Also, monitoring the modes of nanoparticle release and defining their risk impacts can enhance their sustainable utilization by reducing their impacts on plants and the environment. Table 7.2 summarizes the effects caused by different nanomaterials.

7.4 Toxic Effects of Nanomaterials on Soil Ecosystems

7.4.1 Effects on Soil Microbial Communities

Soil microbial communities in soil are comprised of several groups, including fungi, bacteria, actinomycetes, algae, and protozoa. Soil microorganism diversity is a good indicator of soil productivity and fertility. Nanomaterials have shown varied influences on the survival of such soil microbial communities by affecting their activity, community composition, abundance, and diversity. However, the nature of the effects, i.e., negative or positive, depends on the bacterial species subjected to nanomaterials. For instance, soil microorganisms are often negatively affected by metal oxide nanoparticles. Carbon-based nanomaterials show both positive and negative effects. When some carbon-based nanomaterials are added to soil, they are subjected to alter their toxicity due to the reactions that take place with soil organic matter by eliminating their ill effects.

Table 7.2: Negative impacts of nanoparticles on agriculturally important plant species

Crop/plant species	Nanoparticle species	Effects
Wheat (*Triticum aestivum*)	Ag, CuO, ZnO, TiO_2	Decrease of the shoot and root length and fresh biomass, altering the expression proteins involved in primary metabolism and cell defense, reduction of chlorophyll contents increased peroxidase and catalase activities in roots, decreasing of seedling growth, a decrease of yield, lowering phosphorus uptake by plants.
Rice (*Oryza sativa*)	Cu	Reduction in seedling and root growth.
	Ag	Damage the root cell walls and vacuoles of cells changing the morphology and structural features, reduced shoot, and root growth, and damaging the root cells, reducing the total chlorophyll and carotenoid contents.
	CeO_2	Increased the H_2O_2 generation, increase of reactive oxygen species, increase of lipid peroxidation and electrolyte leakage from cells, reducing the quality of rice grains.
	TiO_2	Decrease of biomass, disturbance to the antioxidant defense system, change of metabolite concentration in the plant, suppression of carbohydrate synthesis mechanisms, the elevation of respiration pathways.
	CuO	Decrease of seed germination and seedling growth, severe oxidative burst in plants.
	ZnO	Reduction of root length and number of roots.
	MWCNT	Affecting root and shoot parameters (length and weight).

(Contd.)

Table 7.2: (*Contd.*)

Crop/plant species	Nanoparticle species	Effects
Maize (*Zea maize*)	CuO	Reduction of growth of seedlings.
	Ag	Reduction of germination.
	ZnO	Reduction in root length, altering the root anatomy.
	Al	Reduction in root length and germination.
	TiO$_2$	Delay of seed germination of seeds and decrease of root length, affecting root cell division and causing chromosomal aberrations reduced leaf growth, inhibit root hydraulic conductivity, reduction of transpiration due to formation of aggregates in root walls and reducing the pore diameter, induce oxidative stress and inhibits root growth.
Barley (*Hordeum vulgare*)	CuO	Decrease of shoot and root length and biomass. Increase in H$_2$O$_2$ in plants. Increase the antioxidant enzyme activity.
Chickpea (*Cicer arietinum*)	CuO	Affecting root growth and causing root necrosis, decrease in shoot and root growth and weight, reduction of total chlorophyll content, an increase of reactive oxygen species generation.
Soybean (*Glycine max*)	ZnO	Decrease of root and shoot lengths, surface area and volume, failure in the formation of seeds.
Mung bean (*Vigna radiata*)	Cu	Reduction in seedling and shoot growth.
Tomato (*Lycopersicon esculantum*)	Ag	Reduced root elongation, lowered chlorophyll contents, higher superoxide dismutase activity, and less fruit productivity, decrease in the number of mineral elements, potassium, and iron, appearing deficiency symptoms on leaves.

Plant	Nanomaterial	Effect
Cotton (*Gossypium* spp.)	SiO_2	Decrease of the plant height and shoot and root biomass.
	CeO_2	Decrease of plant height, shoot, and root, decreased contents of nutrient elements (Fe, Ca, Mg, Zn and Na) in roots.
Cucumber (*Cucumis sativus*)	CuO and ZnO	Reduction of plant biomass content, increase of antioxidant enzyme activities, reduction of seed germination, and root elongation.
	MWCNT	Reduction in root length, shoot and root weight, and subsequent cell death.
Pumpkin (*Cucurbita pepo*)	Ag	Reduction in biomass and transpiration.
	MWCNT	Reduction in biomass and transpiration.
	Si	Inhibition of seed germination and transpiration.
Onion (*Allium cepa*)	CuO	Affect the cell cycle and cause chromosomal abnormalities with increased concentrations and duration of exposure.
	Ag	Reduction of root growth and overall plant growth, increase in chromosomal abnormalities in root meristems.
	TiO_2	Reduction of root growth, triggering DNA damage, generation of reactive oxygen species.
	Fullerol ($C_{60}\ OH_2$)	Necrosis, generation of reactive oxygen species, loss of cell membrane integrity.
Lettuce (*Lactuca sativa*)	TiO_2	Reduction of root elongation.
	Al, Zn, MWCNT	Reduction in root growth parameters, decrease of germination of seeds.

Source: Hatami et al. 2016, Rastogi et al. 2017

Thus, in general, carbon-based nanomaterials are less toxic than metal-based ones. Likewise, depending on the nanomaterial species to which the microorganisms are exposed, the level of toxicity would differ (Juárez-Maldonado et al. 2021).

In general, bacterial membranes possess a negative surface charge due to having carboxyl and phosphate groups which facilitates the creation of electrostatic interactions between positive charges of nanomaterials. For instance, metallic nanoparticles adhere to the bacterial membranes in this manner while the released metal ions react with the thiol (R-S-H) groups.

7.4.2 Effects on Nutrient Cycling

Through most of the short-term studies, it has been proven that many negative impacts have been caused on soil microbial communities. The nitrogen cycle is one of the major natural phenomena that enhance agricultural productivity to a greater extent. Affecting the microorganisms associated with the nitrogen cycle can negatively affect the productivity of the crops. Accordingly, metal-based nanomaterials such as Ag, ZnO, TiO$_2$, Ni, Co, CuO and carbon-based nanomaterials such as carbon nanotubes, fullerenes, and graphene have been researched and found to cause negative effects on *Pseudomonas* spp., *Bradirhizobium* spp., *Sinorhizobium* spp. and *Azotobacter* spp. and other heterotrophic bacterial and fungal communities have been reported to be affected by nanoparticles (Wu et al. 2020, Sun et al. 2021).

Nanoparticles can negatively affect such organisms by altering their functioning, for instance, causing elevated levels of unsaturated fatty acids and alteration of the phase transition temperature (SWCNT, *Pseudomonas putida*), affecting the enzyme activity inside the cellular environments of certain bacteria, i.e., phosphorus and potassium solubilizing bacteria. Not only that but plant rhizobacterial communities that are involved in the production of indole acetic acid (IAA) is also affected by certain nanoparticle species. i.e., *Pseudomonas chlororaphis* by ZnO nanoparticles.

Negative effects have been proven through certain indicators such as soil microbial biomass carbon (SMBC), which is a determinant of the balance between soil carbon release and sequestration in the plant-soil-microbial environment. SMBC represents the most active portion of the soil carbon pool. Overall, the effects of nanomaterials are highly dependent upon a variety of soil and plant-related factors, such as pH, the concentration of nanomaterials present, and plant species grown.

7.4.3 Soil Enzyme Activity

Not only the activity of microbes, but also negative effects on soil enzyme activity such as soil protease, catalase, peroxidase, phosphatase, 1,4-ß-N-acetylglucosaminidase, 1,4-ß-glucosidase, cellobiohydrolase, and xylosidase are exerted through the interaction with nanomaterials, i.e., multi-walled carbon nanotubes, TiO_2 and ZnO nanoparticles. Furthermore, Ag and Cu nanomaterials were reported to show significant negative effects on soil enzyme activity (Lin et al. 2022).

7.5 Conclusion and Future Perspectives

Nanomaterials are generated either through natural phenomena or anthropogenic activities, and they exert effects based on their physical and chemical properties in a concentration-dependent manner. These concentrations are highly variable depending on the physical and chemical properties of the nanomaterials and the strength of the physical and chemical properties of the plant species. Toxic effects can be accurately determined through omics studies. Moreover, gene expression studies, high-throughput methods (microarray analysis) and simulation modelling are also novel tools to research the effects of nanomaterials on all living beings. Having this information gap is a barrier to effectively developing nanoformulations with optimized applications such as essential nutrients and other growth enhancers as discussed in the chapter. Therefore, future research should step beyond the basic levels while acquiring the advantage of novel and multidisciplinary involvement to generate a better outcome.

References

Abdel Latef, A.A.H., A.K. Srivastava, M.S.A. El-sadek, M. Kordrostami and L.S.P. Tran. 2018. Titanium dioxide nanoparticles improve growth and enhance tolerance of broad bean plants under saline soil conditions. *Land Degradation & Development*, 29(4), pp. 1065–1073.

Adele, N.C., B.T. Ngwenya, K.V. Heal and J.F.W. Mosselmans. 2018. Soil bacteria override speciation effects on zinc phytotoxicity in zinc-contaminated soils. *Environmental Science & Technology*, 52(6), 3412–3421.

Ahmad, B., A. Shabbir, H. Jaleel, M.M.A. Khan and Y. Sadiq. 2018. Efficacy of titanium dioxide nanoparticles in modulating photosynthesis, peltate glandular trichomes and essential oil production and quality in *Mentha piperita* L. *Current Plant Biology*, 13, pp. 6–15.

Chandra, R. and H. Kang. 2016. Mixed heavy metal stress on photosynthesis, transpiration rate, and chlorophyll content in poplar hybrids. *Forest Science & Technology*, 12(2), pp. 55–61.

García-Sánchez, S., M. Gala and G. Žoldák. 2021. Nanoimpact in plants: Lessons from the transcriptome. *Plants*, 10(4), p. 751.

Hatami, M., K. Kariman and M. Ghorbanpour. 2016. Engineered nanomaterial-mediated changes in the metabolism of terrestrial plants. *Science of the Total Environment*, 571, pp. 275–291.

Hawkesford, M., W. Horst, T. Kichey, H. Lambers, J. Schjoerring, I.S. Møller and P. White. 2012. Functions of macronutrients. In: *Marschner's Mineral Nutrition of Higher Plants* (pp. 135–189). Academic Press.

Judy, J.D., J.K. Kirby, M.J. McLaughlin, T. Cavagnaro and P.M. Bertsch. 2016. Gold nanomaterial uptake from soil is not increased by arbuscular mycorrhizal colonization of *Solanum lycopersicum* (Tomato). *Nanomaterials*, 6(4), 68.

Juárez-Maldonado, A., G. Tortella, O. Rubilar, P. Fincheira and A. Benavides-Mendoza. 2021. Biostimulation and toxicity: The magnitude of the impact of nanomaterials in microorganisms and plants. *Journal of Advanced Research*, 31, pp. 113–126.

Kaňa, R. 2016. Role of ions in the regulation of light-harvesting. *Frontiers in Plant Science*, 7, 1849.

Lin, J., K. Ma, H. Chen, Z. Chen and B. Xing. 2022. Influence of different types of nanomaterials on soil enzyme activity: A global meta-analysis. *Nano Today*, 42, 101345.

Noori, A., J.C. White and L.A. Newman. 2017. Mycorrhizal fungi influence on silver uptake and membrane protein gene expression following silver nanoparticle exposure. *Journal of Nanoparticle Research*, 19(2), p. 66.

Pott, D.M., S. Osorio and J.G. Vallarino. 2019. From central to specialized metabolism: An overview of some secondary compounds derived from the primary metabolism for their role in conferring nutritional and organoleptic characteristics to fruit. *Frontiers in Plant Science*, 10, p. 835.

Rastogi, A., M. Zivcak, O. Sytar, H.M. Kalaji, X. He, S. Mbarki and M. Brestic. 2017. Impact of metal and metal oxide nanoparticles on plant: A critical review. *Frontiers in Chemistry*, 5, p. 78.

Sun, W., F. Dou, C. Li, X. Ma and L.Q. Ma. 2021. Impacts of metallic nanoparticles and transformed products on soil health. *Critical Reviews in Environmental Science and Technology*, 51(10), pp. 973–1002.

Wu, F., Y. You, D. Werner, S. Jiao, J. Hu, X. Zhang, Y. Wan, J. Liu, B. Wang and X. Wang. 2020. Carbon nanomaterials affect carbon cycle-related functions of the soil microbial community and the coupling of nutrient cycles. *Journal of Hazardous Materials*, 390, p. 122144.

Characterization Techniques of Nanomaterials

In nanotechnology, materials characterization is an integral part of research work as it provides information about the chemical composition, physical nature and different properties at macro, micro and nano levels. Depending on the nature of the material being investigated a suite of techniques is utilized to assess its structure and properties. As we are all aware nanoparticles have a high surface area to volume ratio which makes them unique in their bulk states. The size, shape and structure of nanomaterials are highly dependent on their surface properties such as surfactant additives, the concentration of the reactants and conditions applied to the synthesis. However, to obtain reproducibility concerning their synthesis, the characterization of nanomaterials is utterly important. Therefore, this can simply be defined as the analysis of the composition, structure, and physicochemical properties of a given material. There are several characterization techniques available for nanomaterials and this chapter will elaborate on the most frequently used techniques of nanomaterial characterization.

Generally, the characterization techniques utilized typically for nanomaterials can be grouped as spectroscopic methods and imaging methods. Imaging methods are the first choice in most nano-related research where it gives a general overview of particle morphology. Microscopic methods such as electron microscopy are one of the oldest techniques that are still applied. Electron microscopy is a catch-all method that delivers an open view of the material. Electron microscopy consists of two classes Scanning Electron Microscopy (SEM) and Transmission Electron Microscopy (TEM).

8.1 Scanning Electron Microscopy (SEM)

The scanning electron microscopy is one of the most routinely used instruments for the characterization of nanomaterials. In this instrument, an electron beam is emitted from a heated filament, which is commonly made of tungsten by applying a high voltage. The electron beam is accelerated towards the sample by applying an electric field and magnetic lenses to focus it. When primary electrons hit the sample, they give a fraction of their energy to electrons in the sample and it causes them to emit secondary electrons with lower energy. These secondary electrons are collected by a detector and then converted into a voltage, amplified and finally, the image is constructed. The sample must be conducted or covered with a thin metal coating to avoid electrical charging on the surface and the scanning is taking place in a vacuum. In addition to secondary electrons, there are high-energy electrons, originating in the beam which are back-scattered from the sample. These inelastically scattered electrons can be used to detect contrast areas with different chemical compositions. Scanning electron microscopy is used to study morphology or shape, size and internal structures of materials. Also, the interaction of the electron beam with the material emits another form of electron signal known as backscattered electrons. Therefore, detectors in scanning electron microscopes collect both secondary electrons and backscattered electrons to generate images. Backscattered electrons are strongly dependent upon the atomic numbers where the regions with higher atomic numbers in the nanomaterial will appear brighter compared to other areas. Hence, images formed using backscattered electrons can be applied for compositional contrast imaging. Mostly secondary electrons are released close to the surface of the nanomaterial. Also, the geometrical arrangement of the nanomaterial surface determines the degree of secondary electrons thus released. Because these images were generated using secondary electrons they can be used in topographic contrast imaging (Kalantar-zadeh et al. 2008, Kumar et al. 2019).

8.2   Transmission Electron Microscopy (TEM)

In transmission electron microscopy, a beam of focused high-energy electrons is transmitted through a thin sample to reveal information about its morphology, crystallography, particle size distribution, and elemental composition. Transmission Electron Microscopy is capable of providing atomic-resolution lattice images. Transmission electron microscopy can create both electron microscope images (information in real space) and diffraction patterns (information in reciprocal space) for the same region by adjusting the strength of the magnetic lenses. In addition, techniques such as electron energy loss spectroscopy (EELS) associated with transmission electron microscopy can be used to characterize inelastic interactions such as phonon excitations, inter- and intra-band transitions, inner shell ionizations etc. In transmission electron microscopy the most important step is the sample preparation where the sample should be thin for electrons to pass through the materials. The ideal thickness of a given material is based on the technique and material thus utilized. Generally, for transmission electron microscopic applications the thickness of the material should be less than 200 nm (Kalantar-zadeh et al. 2008, Kumar et al. 2019).

8.3   Atomic Force Microscopy (AFM)

Atomic force microscopy is utilized to acquire atomic scale three-dimensional images of surfaces, measuring the attractive or repulsive forces between the scanning probe cantilever and the sample surface. This technique uses a laser beam deflection system, where a laser is reflected from the back of the reflective atomic force microscopic lever and onto a position-sensitive detector. The contact mode where the tip scans the sample in close contact with the surface is the common mode used in atomic force microscopic imaging. In non-contact mode, the tip hovers 50–150 A above the sample surface. Attractive van der Waals forces acting between the tip and the sample are detected and topographic images are constructed by scanning the tip above the surface. Since the attractive forces from the sample are substantially weaker than the forces used by contact mode, the tip must be given a small oscillation. Therefore, AC detection methods can be used to detect the small forces between the tip and the sample by measuring the change in amplitude, phase, or frequency of the oscillating cantilever in response to force gradients from the sample. The tapping mode is a key advance in atomic force microscopy. This potent technique allows high-resolution topographic imaging of sample surfaces that are easily damaged, loosely hold to their substrate, or difficult to image by other techniques. Tapping mode overcomes problems associated with friction, adhesion, electrostatic forces, and other difficulties that plague conventional atomic force microscopic scanning methods by alternately placing the tip in contact with the surface to provide high resolution and then lifting the tip off the surface to avoid dragging the tip across the surface. Tapping mode imaging is implemented in ambient air by oscillating the cantilever assembly at or near the cantilever's resonant frequency using a piezoelectric crystal (Kalantar-zadeh et al. 2008, Kumar et al. 2019).

8.4 Powder X-ray Diffraction (PXRD)

Powder X-ray diffraction is one of the commonly used techniques for the structural and crystal chemical study of crystalline materials. Each crystalline solid has a unique characteristic X-ray diffraction pattern, which may be used as a "fingerprint" for its identification. The basis of X-ray diffraction is Bragg's law, which gives the conditions for diffraction when a monochromatic beam of X-rays is directed onto a crystal.

$$\lambda = 2d_{hkl}\sin\theta$$

Where,

λ - X-ray wavelength

d_{hkl} - The perpendicular distance between the adjacent planes having the Miller indices *hkl*

θ - Bragg angle, the angle between the incident beam of X-rays and the *hkl* plane

For instance, polymers are made up of large chains of monomer units and those polymer chains are arranged themselves in a variety of ways. They can be crystalline, microcrystalline or amorphous and more frequently a mixture of all three forms. Powder X-ray diffraction

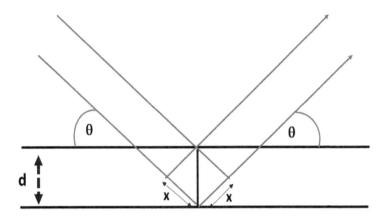

Figure 8.1: Reflection of X-rays from two planes of atoms in a solid.

is a primary technique to determine the degree of crystallinity in polymers. In addition, powder X-ray diffraction also can be used to determine crystal structure (using positions of crystalline reflections and their intensities), crystallite size (Scherer equation) crystalline orientation (Hermans orientation functions) of polymers and crystalline materials. In addition, the degree of intercalation/ exfoliation of layered clays with polymers can be determined by the powder X-ray diffraction technique referring to the basal reflection of layered material as shown in Figure 8.1 (Kalantar-zadeh et al. 2008, Kumar et al. 2019).

8.5   Fourier Transform Infra-Red Spectroscopy (FTIR)

Fourier transform infra-red spectroscopy technique is related to the vibrational frequencies of atoms in a molecule or a solid. Therefore, infrared spectroscopy is employed to identify the type of bond between two or more atoms and consequently to identify functional groups. The peaks in the spectrum correspond to the energies of vibrational transitions within the sample. Therefore, an infrared spectrum is characteristic of a particular compound and provides information about its functional groups, molecular geometry and inter- and intra-molecular interactions.

In fourier transform infra-red spectroscopy, the absorption of light of certain energies that correspond to the vibrational excitation of the molecule is measured. When the vibrational excitation energy of target molecules or compounds of interest lies within the range of 10^{13}–10^{14} Hz it corresponds to the vibrational transitions of the functional groups attached to the nanomaterial surface. This can be used to analyze a compound both qualitatively and observed by infrared spectroscopy, both qualitatively and quantitatively. The energy associated with absorbed infrared radiation is converted into different types of vibrational motions such as stretching (both symmetrical and asymmetrical), scissoring, rocking, wagging and twisting. For a linear molecule with n atoms, there are $3n-5$ vibration modes and if the molecule is non-linear there will be $3n-6$ modes. Not all of these modes of vibration can be excited by infrared radiation. There are selection rules that govern the ability of a molecule to be detected by infrared spectroscopy. For a particular transition to be infra-red active, the vibration must lead to a change in the dipole moment of the molecule.

Infra-red spectroscopy is also widely used to characterize the attachment of the organic ligand to inorganic nanoparticles and it has been widely used to study the bonding interactions between polymers and inorganic fillers in polymer composites. However, inorganic materials are less commonly characterized using infrared radiation since heavy atoms show vibrational transitions in the far IR region. The spectra can be measured not only in transmission/absorption mode but also in attenuated total reflection (ATR) mode, which has been widely used during the last two decades. The main advantage of the ATR mode is that no sample preparation is required and powders, films, gels, and even polymer solutions or dispersions can be characterized fast and easy (Kalantar-zadeh et al. 2008, Kumar et al. 2019).

8.6   Thermal Analysis

8.6.1   Thermo-Gravimetric Analysis (TGA)

The basic principle in thermo-gravimetric analysis is to measure the changes in the mass of a sample as a function of temperature. This is in principle a simple measurement but is an important and powerful tool in solid state chemistry and material science. Thermo-gravimetric analysis can be used to extract mass changes, temperature stability, oxidation and reduction behavior, filler content, moisture content, decomposition compositional analysis of composites etc. Furthermore, thermo-gravimetric analysis measurements are extremely useful in selecting experimental conditions for Differential Scanning Calorimetry (DSC) experiments and for interpreting results (Akash and Rehman 2020).

8.6.2   Differential Scanning Calorimetry (DSC)

Differential scanning calorimetry measures the heat flows associated with transitions in materials as a function of time and temperature. It determines the transition temperatures, melting and crystallization, and heat capacity of a given material. Differential scanning calorimetry uses an inert reference material and a sample. The temperature of the sample is kept identical to the reference material during heating and if a change in temperature between the sample and the reference is detected, the input power to the sample is changed to maintain a constant temperature. In this way, the heat input (output) to (from) the sample during a phase transition can be determined. Differential scanning calorimetry measurements can be used to obtain specific heat, melting crystallization behavior, glass transition temperatures, polymorphism, cross-linking reactions, solid-solid transition etc. The glass transition temperature (Tg) is one of the most important thermos-physical properties of amorphous polymers. This event occurs due to its amorphous nature which facilitates the macromolecular motions (translational) of polymer chains. This is an important event as changes in molecular mobility at Tg causes significant changes in the physical and reactive properties of the polymer (Akash and Rehman 2020).

8.6.3 Dynamic Mechanical Analysis (DMA)

Dynamic mechanical analysis is a technique where a small deformation is applied to a sample as an oscillating force and the mechanical properties of the material are measured as a function of temperature, frequency and strain. Dynamic mechanical analysis measures stiffness and damping as modulus and tan delta. As sinusoidal force is used in-phase component is considered the storage modulus whereas, out of phase component is the loss modulus. The storage modulus measures the sample's elastic behavior and the ratio of the loss to the storage is the tan delta which is called damping and measures the energy dissipation of the material. Dynamic Mechanical analysis provides data on the composition and structure of polymer blends (miscibility), the glass transition temperature of highly cross-linked, amorphous or semi-crystalline polymers and composites, curing/post-curing, aging, creep and relaxation stress and strain sweeps etc. (Menard and Menard 2020).

8.7 Conclusion and Future Perspectives

Increased demand for nanotechnology for several industrial applications has led to it being utilized by different types of nanomaterials. Therefore, characterization is of utter importance specifically to analyze and understand the specific properties of synthesized nanomaterials. There are numerous techniques which have been introduced in characterizing nanomaterials. Microscopic-based approaches such as electron microscopy, thermal analysis methods, and spectroscopic methods allow for precisely identifying nanomaterials' properties and behavior. However, nanomaterial type and their composition, dimensions, and the environment in which the study is performed affect the characterization techniques. Therefore, every method of characterization has its own merits and demerits and a proper selection of characterization methods allows the precise identification of their properties. In addition to the above-mentioned techniques energy electron diffraction, nuclear reaction analysis, fast neutron analysis, and nuclear reaction analysis are used for chemical analysis of nanomaterials.

References

Akash, M.S.H. and K. Rehman. 2020. Thermo gravimetric analysis. In: *Essentials of Pharmaceutical Analysis* (pp. 215–222). Springer, Singapore.

Kalantar-zadeh, K. and B. Fry. 2008. Characterization techniques for nanomaterials. *Nanotechnology-Enabled Sensors*, pp. 211–281.

Kumar, P.S., K.G. Pavithra and M. Naushad. 2019. Characterization techniques for nanomaterials. In: *Nanomaterials for Solar Cell Applications* (pp. 97–124). Elsevier.

Menard, K.P. and N.R. Menard. 2020. *Dynamic Mechanical Analysis*. CRC Press.

Index